Remote Sensing and Cognition

Human Factors in Image Interpretation

Remote Sensing and Cognition
Human Factors in Image Interpretation

Edited by
Raechel A. White, Arzu Çöltekin, and
Robert R. Hoffman

CRC Press
Taylor & Francis Group
Boca Raton London New York

CRC Press is an imprint of the
Taylor & Francis Group, an **informa** business

CRC Press
Taylor & Francis Group
6000 Broken Sound Parkway NW, Suite 300
Boca Raton, FL 33487-2742

First issued in paperback 2020

© 2018 by Taylor & Francis Group, LLC
CRC Press is an imprint of Taylor & Francis Group, an Informa business

No claim to original U.S. Government works

ISBN-13: 978-0-367-57178-8 (pbk)
ISBN-13: 978-1-4987-8156-5 (hbk)

Library of Congress Cataloging-in-Publication Data

Names: White, Raechel A., editor. | Coltekin, Arzu, editor. | Hoffman, Robert
R., editor.
Title: Remote sensing and cognition : human factors in image interpretation /
edited by Raechel A. White, Arzu Coltekin, and Robert R. Hoffman.
Description: Boca Raton, FL : Taylor & Francis, 2018. | Includes
bibliographical references.
Identifiers: LCCN 2017055871 | ISBN 9781498781565 (hardback : acid-free
paper)
Subjects: LCSH: Remote-sensing. | Geographical perception. | Cognition.
Classification: LCC G70.4 .R4555 2018 | DDC 910/.019--dc23
LC record available at https://lccn.loc.gov/2017055871

Visit the Taylor & Francis Web site at
http://www.taylorandfrancis.com

and the CRC Press Web site at
http://www.crcpress.com

Dedicated to my son, Brett Bianchetti.

Raechel White

Dedicated to my parents, who were my first teachers, and all others who
generously shared their knowledge along the way. Thank you.

Arzu Çöltekin

I would like to dedicate this book to Paul J. Feltovich, to acknowledge his
many contributions to cognitive science and the field of expertise studies.
In particular, his theories of complex reasoning and expert perception
have had a huge impact. This makes me especially privileged to say that
he has been, and remains, a dear friend and esteemed colleague.

Robert R. Hoffman

Contents

Preface

CRC Press, the publisher of *Interpreting Remote Sensing Imagery: Human Factors* (2001), expressed interest in a second edition a few times over the ensuing years. But one of the editors of that volume, R. Hoffman, had shifted his research focus from the psychology of perception to the study of expertise in weather forecasting. The other editor, A. Markman, had shifted his research focus from the psychology of decision making to organizational psychology.

Meanwhile, the field of photogrammetric engineering morphed into remote sensing and then further morphed into what is now called geospatial analysis. In the interim, in the field of expertise studies, models of expert decision making had been maturing. In particular, the Data/Frame model of sensemaking by Gary Klein and his colleagues (Klein, Moon, & Hoffman, 2006) had been gaining traction in the literature on judgment and decision making.

Serious consideration of the second volume of this work began after interactions between Hoffman and faculty at The Pennsylvania State University, and more specifically, the work on geospatial intelligence sensemaking by Todd Bacastow and his colleagues at The Pennsylvania State University. The correspondence that ensued began with a focus on models of sensemaking, and in particular, expertise in remote sensing image interpretation. The discussions soon led to the recognition that the human factors of remote sensing imagery did indeed need a new synthesis, and the scope and reach of research had grown considerably.

A series of emails and teleconferences in April 2014 generated names of a new generation of researchers and scholars whose work fell at the intersection of psychology (or human factors) and remote sensing. Todd Bacastow and Dennis Bellafiore of the Pennsylvania State University were the connection-makers, hooking the three editors of this volume together. Thus, this book was born. Raechel White's own PhD work at The Pennsylvania State University, considering the role of expertise in remote sensing image interpretation, brought her onto the project early on as a contributor and editor. As the work progressed, she recruited Arzu Çöltekin, first as a contributor due to her expertise in eye-tracking and photogrammetric imaging and later as a co-editor for the volume. Together, White, Çöltekin, and Hoffman have overseen the development of this book as a team since January, 2015.

We wanted this new volume to cover topics and research that were new and thus, not simply to be an update of the 2001 book. The thematic topics in the 2001 volume were physiography and landform analysis via expert systems, sensor fusion, and human factors of remote sensing in meteorology. These were all new, cross-disciplinary topics at that time, involving remote sensing with computer science and experimental psychology methodology.

While these are no longer "new topics," the ideas and methods described in the 2001 book remain valuable today. Thus, we see this new book as adding to, rather than merely "updating," the 2001 book. That said, the thematic topics here also involve the merger of psychological research methods with issues in remote sensing, including map interpretation, satellite image interpretation, and analysis training.

Raechel White
Arzu Çöltekin
Robert Hoffman

Reference

Klein, G., Moon, B., & Hoffman, R. R. (2006). Making sense of sensemaking 2: A macrocognitive model. *IEEE Intelligent Systems*, 21(5), 88–92.

About the Authors

 **Raechel White, PhD** is an early-career scientist whose work addresses the role of remote sensing imagery in deriving information about our natural environment. Her work is driven by a keen interest in information extraction, the vertical viewpoint, and technology. Through the use of cognitive systems engineering approaches, Dr. White develops models of human interpretation practices. Her work contributes to our understanding of how remotely sensed imagery can be a source of information for geovisual analytic systems.

Dr. White has been active within a number of professional organizations championing the need for more focus on the critical role that human reasoning plays in information extraction. Her work in improving student learning experiences in remote sensing has been recognized by the International Society for Photogrammetry and Remote Sensing.

Dr. White's PhD is in geography from Pennsylvania State University, where she received a number of scholarships, including from the National Science Foundation and the National Geospatial Intelligence Agency, based on her innovative work in cognitive aspects of change detection. Following her PhD, she took up an assistant professorial position at Michigan State University, where she teaches geographic object-based image analysis and cartography.

 **Arzu Çöltekin, PhD** is an established scientist known for her interdisciplinary work in scientific visualization and human–computer interaction in the context of geography. She examines our understanding of the human visual system (vision science: optics, perception, cognition) and links it to how we design, and interact with, visualizations through a lens of attention and memory. Thus, through empirical and technological methods, she contributes to improving the design and use of scientific visualizations, as well as to our understanding of how humans process visual and spatial information.

Çöltekin is the chair of the Geovisualization, Augmented, and Virtual Reality working group that operates within the International Society of Photogrammetry and Remote Sensing and a research affiliate with Harvard University's Seamless Astronomy Group. She has led many international workshops, co-organized conferences, and edited journal special issues. She has received many national and international scholarships and awards through her early career and obtained prestigious research grants as a PI and as a co-PI in later years (e.g., from the Swiss National Science Foundation and the Royal Society in the UK), and, most recently, was selected for the Google Faculty Research Award as a PI for her proposal on visual complexity.

Çöltekin's PhD is in photogrammetry and remote sensing with a minor in cartography and geoinformatics from the Helsinki University of Technology (now Aalto University), where she has also worked as a teaching fellow. Following a postdoctoral position at the Media Lab of the University of Art and Design Helsinki, she joined the Department of Geography at the University of Zurich in Switzerland. All of her publications are available for download at http://www.geo.uzh.ch/~arzu/publications/.

 **Robert R. Hoffman, PhD** is a recognized world leader in cognitive systems engineering, expertise studies, and human-centered computing. He is a fellow of the Association for Psychological Science, fellow of the Human Factors and Ergonomics Society, senior member of the Association for the Advancement of Artificial Intelligence, senior member of the IEEE, and a Fulbright scholar. He has been principal investigator, co-principal investigator, principal scientist, senior research scientist, principal author, or principal subcontractor on over 60 grants and contracts totaling nearly $12 million. He has led efforts including large, multi-partner, multi-year grant collaborations, contracted alliances of university and private sector partners, and multi-university research initiatives. His PhD is in experimental psychology from the University of Cincinnati, where he received McMicken Scholar, Psi Chi, and Delta Tau Kappa honors. Following a postdoctoral associateship at the Center for Research on Human Learning at the University of Minnesota, he joined the faculty of the Institute for Advanced Psychological Studies at Adelphi University. He pioneered the study of methods for eliciting the knowledge of domain experts. Hoffman has been recognized internationally in psychology, remote sensing, human factors engineering, and artificial intelligence for his research on the psychology of expertise, the methodology of cognitive task analysis, human-centered computing issues for intelligent systems technology, and the design of cognitive work systems. Hoffman is a co-editor for the Department on Human-Centered Computing in *IEEE: Intelligent Systems.* He is editor for the book series "Expertise: Research and Applications." He was a co-founder of *The Journal of Cognitive Engineering and Decision Making.* A full curriculum vitae and all of his publications are available for download at www.ihmc.us/groups/rhoffman/.

Contributors

Jagannath Aryal
Discipline of Geography and Spatial
  Sciences
University of Tasmania
Hobart, Australia

Todd S. Bacastow
Dutton e-Education Institute
The Pennsylvania State University
University Park, PA

Raffaella Balzarini
Pervasive Team
Inria Grenoble Rhône-Alpes
Montbonnot-Saint-Martin, France

Dennis Bellafiore
Dutton e-Education Institute
The Pennsylvania State University
University Park, PA

Susan P. Coster
Raytheon, Inc.
Waltham, MA

Stephen P. Handwerk
Dutton e-Education Institute
The Pennsylvania State University
University Park, PA

Pyry Kettunen
Department of Geoinformatics and
  Cartography
Finnish Geospatial Research
  Institute
Helsinki, Finland

Lester C. Loschky
Department of Psychological
  Sciences
Kansas State University
Manhattan, KS

Arko Lucieer
Discipline of Geography and Spatial
  Sciences
University of Tasmania
Hobart, Australia

Nadine Mandran
Laboratoire d'Informatique de
  Grenoble
Université Grenoble Alpes
Saint-Martin-d'Hères, France

Robert Musk
Resources and Planning
Sustainable Timber Tasmania
Hobart, Australia

Jon Osborn
Discipline of Geography and Spatial
  Sciences
University of Tasmania
Hobart, Australia

Stanislav Popelka
Department of Geoinformatics
Palacký University Olomouc
Olomouc, Czech Republic

Sachit Rajbhandari
Discipline of Geography and Spatial
  Sciences
University of Tasmania
Hobart, Australia

Ryan V. Ringer
Department of Psychological
    Sciences
Kansas State University
Manhattan, KS

Laura D. Strater
Raytheon, Inc.
Waltham, MA

Gregory Thomas
Dutton e-Education Institute
The Pennsylvania State University
University Park, PA

1

Cognitive and Perceptual Processes in Remote Sensing Image Interpretation

Robert R. Hoffman

CONTENTS

KEYWORDS: *expertise, perception, reasoning, image interpretation, expert systems, knowledge elicitation, Critical Decision Method, knowledge representation, Concept Maps, terrain analysis, recognition-primed decision making, Data/Frame model of sensemaking, naturalistic decision making*

1.1 Introduction

It has been over 17 years since the idea was born of a book on psychological factors in remote sensing (Hoffman & Markman, 2001). Since then, there have been astounding advances in the technology of remote sensing: new sensing systems, new data types and products, new visualizations, new software systems to support image processing, and even artificial intelligence systems for scene and pattern recognition. Likewise, there have been advances on the psychology side. These include advances in our understanding of perceptual learning and reasoning processes that are crucial in image analysis. The purpose of this chapter is to encapsulate the state of the art regarding expert reasoning and perception. First comes some background.

1.2 That Was Then

Psychological considerations of display design, analytical method, and image interpretation were at the forefront as the field of photogrammetic engineering was morphing into the field of remote sensing. The findings and ideas from decades of research on cartographic communication were applied in the consideration of the use of color coding in remote sensing imagery (e.g., Hoffman & Conway, 1990). Researchers were beginning to conduct experiments in which expert remote sensing scientists were presented with various image interpretation tasks to reveal differences between what they could perceive and what novices could (and could not) perceive (e.g., Hoffman, Markman, & Carnahan, 2001; Lowe, 2001). A number of efforts were underway at that time to generate "expert systems" for image interpretation, and this too meant that there was a role for psychologists in the remote sensing enterprise. Specifically, the inference rules and knowledge bases of expert systems had to be created from knowledge elicitation interviews with experts. An entire field was emerging in parallel to remote sensing—the field of expert knowledge elicitation. And that methodology was being applied in the study of expert aerial photo interpreters (Argialas & Milaresis, 2001; Hoffman, 1984, 1985, 1987; Hoffman & Pike, 1995). In hindsight, only the most basic of psychological principles were discussed at that time, but the invocation of those was important.

First, it was clear that color coding of meaning was a matter not to be taken lightly. The common "rainbow code" had issues. Schemes of all sorts for depicting infra-red and other wavelength imagery were being created by the developers of the sensing and display technology, with little or no input about visualization design from human factors psychologists. We know that all encoding schemes have issues, or potential issues, of meaning, interpretation, and misinterpretation.

Second, it was clear that experts possess an incredible wealth of knowledge. The struggle to develop good knowledge elicitation methods—by adapting methods from the psychology laboratory and methods of structured interviewing—was motivated, in part, by the fact that it took a long time for expert system developers to create their sets of concepts and inference rules, more time even than it took to program the expert system. This triggered a key question: Exactly how much do experts know? Estimates were in the tens of thousands of concepts, chunks, patterns, or propositions. This was a daunting consideration, one that contributed to the emergence of the field of expertise studies (see Ericsson, Hoffman, Kozbelt, & Williams, 2018; Hoffman, 1987).

Third, it was clear that experts perceive patterns, not cues (e.g., Lowe, 2001; Mogil, 2001). Even though cues contribute to the perception of patterns, patterns can be more than the aggregation of cues, because they reference the interpreter's knowledge—as it has been said, in a "top-down" manner.

Hoffman (1990) distinguished between remote *sensing* (which is technology-oriented) and remote *perceiving* (which is meaning-oriented). While expert remote sensing image interpreters might *see* images and cues, what they *perceive* is geobiological dynamics. While the novice might see what appears to be a cigar-shaped hill, the expert perceives the eons-long glacial dynamics that led to the formation of a field of drumlins. While the novice sees a network of streams leading to a pond in the midst of a heavily vegetated area, the expert perceives the underlying flat-lying limestone, implying stagnancy that threatens bacterial infections for individuals who might go there. What seems to novices to be a chain of inferences is, in fact, an act of "direct perception" to the expert. Experts at remote sensing image interpretation seem able to perceive the invisible (Klein & Hoffman, 1992).

These key ideas lead directly to recent advances in cognitive science. What more has been learned about expert knowledge? How does cognitive science now theorize about expert reasoning?

1.3 This Is Now

1.3.1 Eliciting Knowledge

The daunting challenge of capturing expert knowledge has always been understood not as merely an academic matter but as a very practical one. All professions rely on cadres of experts who "keep the light bulbs burning." Knowledge management is another emergent discipline that keys into the topic at hand. Knowledge management has the goal of capturing expert knowledge so that it might be integrated into training programs to help build the next generation of experts (Ackerman, Pipek, & Wulf, 2003; Becerra-Fernandez, Gonzalez, & Sabherwal, 2004; Hoffman et al., 2014). Methods of knowledge elicitation are now well understood, including strengths, appropriate uses, and matches to domains and application areas. A number of scholarly works describe knowledge elicitation and cognitive task analysis methods in detail (e.g., Burton, Shadbolt, Rugg, & Hedgecock, 1990; Crandall, Klein, & Hoffman, 2006; Schraagen, Chipman, & Shalin, 2000; Vicente, 1999).

One particularly useful method is called the critical decision method (CDM) (Klein, Calderwood, & MacGregor, 1989). The CDM was created and refined during the era of expert systems and the rising interest in expertise studies. Rather than asking experts "Tell me everything you know about x," the CDM works by scaffolding the expert in the recounting of previously lived "tough cases." These are cases where the expert's knowledge and skills were stretched. Experts often have vivid memories of such cases and can recount them in detail. This includes the creation of a timeline for the incident, with annotations of when crucial observations or key decisions were

TABLE 1.1

Example Knowledge Elicitation Probe Questions

Information	What information did you need or use to make this judgment?
	Where did you get this information?
	What did you do with this information?
Mental modeling	As you went through the process of understanding this situation, did you build a conceptual of the problem scenario?
	Did you try to imagine the important causal events and principles?
	Did you make a spatial picture in your head?
	Can you draw me an example of what it looks like?
Knowledge	In what ways did you rely on your knowledge about this kind of case?
	How did this case relate to typical cases you have encountered?
	How did you use your knowledge of typical patterns?
Guidance	At what point did you look at any sort of guidance?
	Why did you look at the guidance?
	How did you know if you could trust the guidance?
	How do you know which guidance to trust?
	Did you need to modify the guidance?
	When you modify the guidance, what information do you need or use?
	What do you do with that information?
"What if" probes	"At this point in the incident, what if it had been a novice present, rather than someone with your level of proficiency?
	Would they have noticed Y?
	Would they have known to do X?"

Source: Adapted from Crandall, B., Klein, G., and Hoffman, R.R. *Working Minds: A Practitioner's Guide to Cognitive Task Analysis.* MIT Press, Cambridge, MA, 2006 and Hoffman, R.R., Crandall, B., and Klein, G. Protocols for Cognitive Task Analysis. Report, Florida Institute for Human and Machine Cognition, Pensacola FL, 2008. Retrieved from www.dtic.mil/cgi-bin/GetTRDoc?AD=ADA475456.

made. Probe questions are posed with respect to the key points on the timeline to elicit information about strategies and the basis for decisions and the perceptual cues on which the decision maker relies. Example knowledge elicitation probe questions are presented in Table 1.1.

The CDM has a significant track record of success at eliciting expert knowledge in various domains, including the domain of remote sensing. An example of this is provided in the next section.

1.3.2 Representing Knowledge

The representation of the knowledge of experts in any domain, including remote sensing, has been facilitated significantly by the use of Concept Maps. The technique of creating Concept Maps in knowledge elicitation interviews has been used to support knowledge sharing and knowledge preservation in diverse domains across the world (see Moon, Hoffman, Cañas, & Novak, 2011; Novak, 1998). Concept Maps are meaningful diagrams in which nodes represent concepts and linking lines express relations between concepts. Using the CmapTools freeware developed at the Institute for Human & Machine

Cognition (IHMC) (www.ihmc.us/cmaptools/), it is possible to append any form of digital media onto the nodes in a Concept Map.

An example application in remote sensing was the System To Organize Representations in Meteorology (STORM) project (Hoffman, Coffey, Ford, & Novak, 2006). This project involved capturing the knowledge and reasoning of expert weather forecasters, and it included an effort to capture knowledge specifically about the interpretation of weather radar imagery. About a dozen Concept Maps were made about various radar products, the kinds of meaningful patterns that could be seen in the images, and so on. The project focused on expert rules for forecasting severe weather in the Gulf Coast region of the United States, but it was suggestive of how Concept Maps could be used to capture the knowledge of remote sensing scientists. The next project took this idea to an entirely new scale.

The project was called The Representation of Conceptual Knowledge in Terrain Analysis (ROCK) (Eccles, Feltovich, & Hoffman, 2004; Hoffman, 2007). It was an instantiation of the Terrain Analysis Database, which had been created at the U.S. Army Corps of Engineers (Hoffman, 1984). This database consisted of over 1500 assertions concerning terrain, terrain forms, landforms, rock types, and so on, as these are depicted in aerial photos. The database was created by knowledge elicitation interviews, including the CDM, with the participation of expert image analysts at the Topographic Engineering Laboratory (now the Topographic Engineering Center). As a traditional text document, the Terrain Analysis Database was of limited usability.

The process of creating ROCK involved mapping all of the propositions. For example, the database assertion "Dome granite can be massive domes with widely spaced joints" consists of two propositions: "Dome granite can be massive domes" and "Massive domes have widely spaced joints." The database entry for thickly interbedded sandstone, limestone, and shale included these assertions:

- Sharp, thick, angular, dissected ridges
- Tonal bands to rock
- Flat valleys
- Faults, folds, fractures, trellis drainage on slopes and ridges
- Dendritic drainage in valleys
- Sinkholes
- Quarries in limestone valleys
- Forested ridges and slopes in sandstone or limestone

Some of these assertions actually consist of multiple propositions (e.g., ridges can be thick, ridges can be sharp, etc.). All the individual propositions were extracted from the database assertions and expressed as concept-link-concept triples in the Concept Maps. The completed ROCK knowledge model consisted of 150 Concept Maps containing 3341 concepts, 1634 relational links, and 3352 propositions (average of 22 propositions per Concept Map). Figure 1.1 presents the Concept Map for terminal moraine. Figure 1.2 presents the Concept Map for tilted interbedded sedimentary rock with shale predominating.

After the Concept Maps were drafted, they were reviewed by an expert aerial photo interpreter, who offered suggestions and corrections. Actually, most of the expert's comments were verifications, and most of the suggestions for changes were wordsmithing. This result verified the finding that experts can show high levels of agreement.

In the upper left corner of Figures 1.1 and 1.2, one can see what appears to be a stack of nodes. This is the ROCK "navigator," which shows the user where they are in the overall knowledge model. Each level, moving upward in the stack, takes one to a Concept Map that integrates all the Concept Maps that fall under it in the knowledge hierarchy. The navigator makes it possible to go from any given Concept Map to any other in as few as two clicks while never losing the big picture of where they are in the overall knowledge model.

Additional materials were available for use in creating ROCK. These included topographic maps, representative stereo aerial photographs, and photo interpretation keys. These were integrated into the ROCK Concept Maps through the mechanism for linking media to concept nodes. The entire Army Field Manual for Terrain Analysis (Department of the Army, 1990) was carved into its meaningful pieces, and these were appended to the Concept Maps. As can be seen in Figures 1.1 and 1.2, icons beneath some nodes could indicate a linkage to some other Concept Maps (e.g., the node for "Glacial Landform" in the terminal moraine Concept Maps links to a Concept Map that is just about the various glacial landforms). Other icons indicate linkage to text documents, stereo aerial image examples, and text documents. Figure 1.3 shows two ROCK Concept Maps: the high-level Concept Map that organizes all the Concept Maps about dunes, and the Concept Map specifically about star dunes. Appended to nodes in the star dunes Concept Map are various resources, showing the appearance of a star dune field from a terrestrial perspective and from a topographic perspective, along with a text piece about trafficability.

The intended immediate application of ROCK was that warfighters who entered a region could "stand on the shoulders of experts" and learn about the terrain in which they would find themselves and its implications for military operations (e.g., trafficability). But the broader application was in the capture of expert knowledge for use in knowledge sharing and training. At the time of its creation, ROCK was the largest model of expert knowledge that had been created, and yet it was felt that the 150 Concept Maps were far from a complete expression of expert knowledge in terrain analysis. Nevertheless, ROCK exemplified a way to capture the knowledge of remote sensing experts. That endeavor became especially salient as more

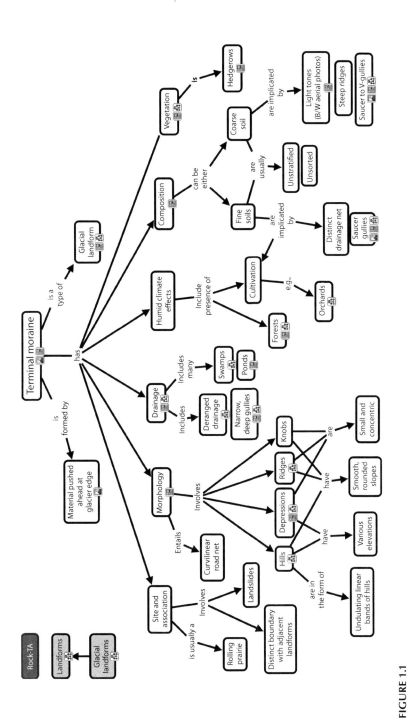

FIGURE 1.1
Concept Map about the appearance of a terminal moraine in aerial photographs.

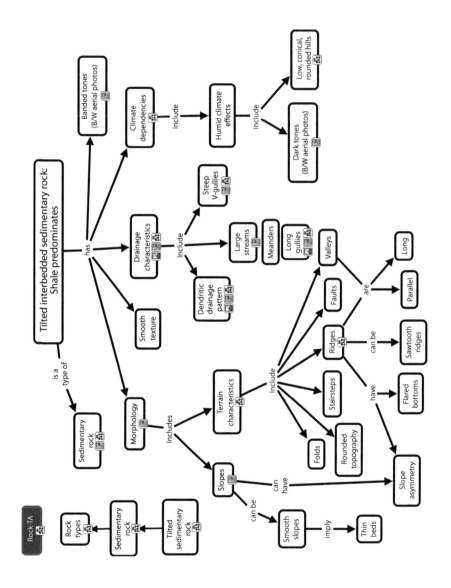

FIGURE 1.2
Concept Map about tilted interbedded sedimentary rock with shale predominating.

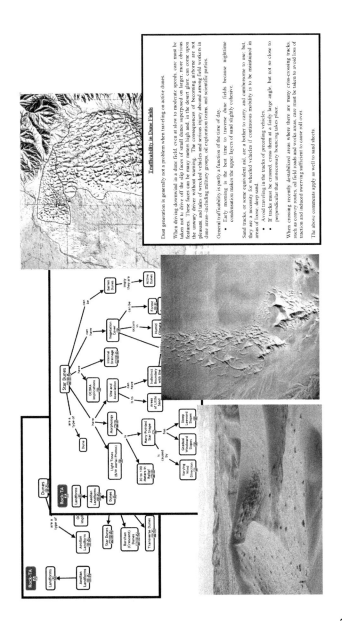

FIGURE 1.3
The ROCK Concept Map about star dunes and some of the appended resources.

and more people expressed concern about the "grey tsunami," that is, the immanent loss of an entire generation of experts of the "boomer" generation (see Hoffman & Hanes, 2003). One of the key findings from the ROCK project was that the "knowledge recovery" process of transforming the Terrain Analysis Data Base (text document) into the navigable Concept Maps was more effortful and time-consuming than the alternative process—representing knowledge by Concept Maps in the first place.

I now turn to a second important area of advance in cognitive science that pertains directly to remote sensing.

1.3.3 Models of Reasoning

As cognitive science emerged in the 1970s, the leading models emphasized the mechanisms of short-term and long-term memory. This emphasis was driven primarily by the research paradigm of "verbal learning," in which college students taking introductory psychology served as subjects in tasks involving recognition, recall, comprehension, and reaction time for such materials as word lists, sentences, and other isolated stimuli. The tasks, from the instructions to final performance, had to be doable within the duration of a college class period.

Discontent with this paradigm began to emerge when it became clear that context and meaning were crucial factors in cognition (Jenkins, 1974; Miller, 1986; Newell, 1973). Cognitive psychology had focused almost entirely on the micro level of analysis and had lost sight of the bigger questions, such as the nature of knowledge and the ways cognition works in the "real world" (Hutchins, 1995).

In parallel, seminal research was being conducted that looked at the cognition of experts. One such study, by Gary Klein and his colleagues (Klein, Calderwood, & Clinton-Corroco, 1986; reproduced as Klein, Calderwood, & Clinton-Corroco, 2010), involved using the Critical Decision Method to investigate the decision making of firefighters. It is important to understand that in this era the predominant model of decision making was "utility analysis," which had emerged from the field of judgment and decision making. In utility analysis, the decision maker lists alternative decisions, judges the likelihood of the various outcomes, and selects a course of action that minimizes risk and maximizes gain. What Klein and his colleagues found was that in domains involving dynamic situations, time pressure, high risk, stress, and uncertainty, decision makers do not engage in utility analysis or the comparison of alternative courses of action. Instead, they size up the situation and seem to know immediately what the best course of action is.

Indeed, as shown by other research, forcing people to engage in utility analysis can interfere with their reasoning. Zakay and Wooler (1984) trained people in utility analysis, and found that their problem solving could proceed effectively using the strategy if there was no time pressure. But if even moderate time pressure was imposed, the strategy was not beneficial. Beach and Lipshitz (1993) analyzed the decision-making protocols of military commanders in terms of the decision analytic model and found that the analysis distorted the actual strategies and reasoning sequences and failed to capture the recognitional aspects of command decision making.

In many domains of expert decision making, such as firefighting, power plant operation, jurisprudence, and design engineering, experts often make decisions through rapid recognition of causal factors and goals, rather than through any explicit process of generating and evaluating solutions (Klein & Hoffman, 1992). It was that finding—the crucial role of recognition in decision making—that led to the Recognition-Primed Decision Making (RPD) model (Klein, 1989). Previous psychological research on decision making and on recognition had studied the two processes separately and had not linked them. Klein proposed a model, depicted in Figure 1.4.

Further research on various domains of expertise (for a review, see Hoffman & Militello, 2008) led to refinements of the RPD model. For example, a decision maker might have latched onto a preferred course of action, but during the implementation, an anomaly might occur. This would lead to a re-assessment of the situation. Additionally, the decision maker would engage in mental simulation in which the anticipated unfolding of the event is imagined, leading to modifications in the action plan. Over the course of a decade of research, the findings from various studies of experts converged on a model called the Data/Frame model of sensemaking (Klein, Moon, & Hoffman, 2006a,b; Klein, Phillips, Rall, & Peluso, 2003). This model is shown in Figure 1.5. Sensemaking is considered to be a conscious and deliberate activity aimed at making sense of events or phenomena (Weick, 1995).

When one looks retrospectively at an instance of reasoning, one can describe it as a causal chain, as in: (1) an initial "frame" (or conceptual model) is formed, (2) it is used to seek confirming and disconfirming evidence, and (3) the frame is elaborated, discarded, and so forth. But before the fact, as a generic process model, all of the sequences are possible. Hence, the Data/Frame model consists entirely of closed loops. The closed loop at the top expresses the notion that the decision maker's initial "frame" for understanding the data is shaped by a few key data points or elements, and at the same time, the frame determines what counts as data. Both the data and the frame are inferred and are not perceptual primitives. Once a frame is established, there are a number of possible paths. The frame might be questioned, based on the observation of an anomaly. Further reasoning might lead to an elaboration of the frame or a search for an entirely different frame. Sensemaking usually proceeds in fits and starts; there can be gaps, distractions, and multitaskings. Beginnings and endings can be anything but clear-cut. Although we can retrospectively describe instances in which reasoning seems to follow a sequence with a clear-cut starting point ("surprise" is often the trigger to problem solving) and an apparent stopping

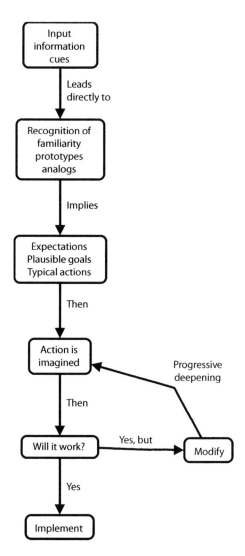

FIGURE 1.4
The initial Recognition-Primed Decision Making model. (Adapted from Calderwood, R., Crandall, B., and Klein, G. Expert and novice fireground command decisions. Report MDA903-85-C-0327, U.S. Army Research Institute, Alexandria, VA, 1987.)

point (a decision is "made"), such causal chains are the exception and not the rule (see Hoffman & Yates, 2005).

It was noted earlier that the research on RPD sometimes revealed a process in which the noticing of an anomaly led the decision maker to reconsider their course of action. The sensemaking process shown in Figure 1.5 is intended to describe what happens in that reframing process. Sensemaking

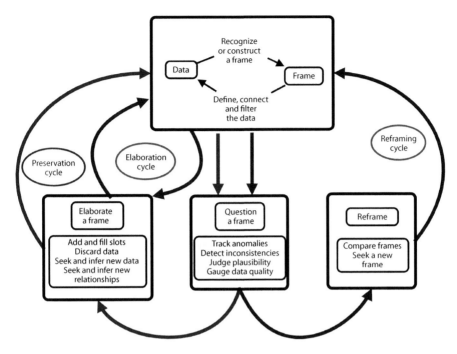

FIGURE 1.5
The Data/Frame model of sensemaking. (Adapted from Klein, G., Phillips, J.K., Rall, E., and Peluso, D.A. A Data/Frame theory of sensemaking. In R. Hoffman (Ed.), *Expertise out of Context*. Erlbaum, Mahwah, NJ, 2003.)

is used to achieve a functional understanding—what to do in a situation—as well as an abstract understanding (Klein et al., 2003a).

Having presented these two models—RPD and Data/Frame—we can see now how they apply to the case of remote sensing image interpretation. Once an individual has achieved high proficiency at image interpretation, they can view an image and directly perceive the key dynamics and patterns, and go from that to some course of action, say, a classification of features or of the entire image. This would be RPD. But for individuals who are less proficient, and for experts who notice some anomaly, the full sensemaking process is engaged. These models provide a far richer description of the reasoning of remote sensing image interpreters than derives from the simple assertion that image interpretation is just abductive inference.

Consider, for example, the weather forecaster who is interpreting radar and satellite images. They do not read off information from images and charts. They inspect the images and charts and form an initial frame about the primary atmospheric dynamics. At first, the frame is the "big picture," based primarily on key qualitative information (e.g., "The jet stream is further south today" or "The low seems to be developing along the Louisiana coast"). This initial frame helps them establish the "forecasting problem of the day." They project their mental model in time and watch for signals indicating that an event might be unfolding or becoming unusual. They continuously refine their frame, and they rely heavily on it to generate a forecast (Hahn, Rall, & Klinger, 2003; Hoffman et al., 2017).

1.4 Conclusions

We now know a great deal about expertise (for reviews, see Ericsson et al., 2018; Hoffman et al., 2014), and indeed, expertise studies has become a field in its own right. Seminal research by Klein and others on the perception and reasoning of domain experts led to the emergence of a new paradigm called Naturalistic Decision Making (NDM). It is regarded as complementary to traditional experimental psychology in that it is a form of naturalism (although much NDM research involves the control and manipulation of variables in experimentation). NDM is distinguished from traditional experimental psychology insofar as it studies how cognition adapts to complexity by looking at such processes as sensemaking, mental model formation, problem detection, and coordination, as these are manifested in professional domains, including remote sensing (Klein et al., 2003b; Schraagen et al., 2008). When an expert is working on a tough problem in context, they often have to deviate from standard procedures and engage in problem solving in which new methods are created on the fly. In complex cognitive work, activities are rarely tasks in the sense of fixed action sequences. Instead, domain practitioners engage in knowledge-driven, context-sensitive choice among activities. Thus, for any given application domain, such as specialist work in remote sensing, there cannot be some single model of the reasoning process. Instead, there needs to be a model patterned after Data/Frame, which integrates multiple possible paths to solutions.

An especially exciting challenge is to better understand how experts can interpret and integrate multiple data types, including data that are dynamic (e.g., satellite loops). Historically, psychology distinguished perception from learning: they were regarded as fundamental but distinct cognitive operations. Although the concept of perceptual learning was proposed in developmental psychology decades ago (Gibson, 1963), the study of perceptual learning requires long-term, longitudinal observation, and this generally falls outside of the traditional experimental psychology paradigm for studying cognition in the academic laboratory. In contrast, the perceptual learning capacity of experts has long been recognized in expertise studies (cf. Klein & Hoffman, 1992). That said, advances in technology have raised the stakes. New data types, new methods for data visualization and display, and new software systems for data analysis and integration all mean that experts not only have to learn how to perceive meaning but have to continuously un-learn and re-learn what to perceive (Hoffman & Fiore, 2007).

The study of macro-level cognition can be seen as the culmination of the interplay of all the paradigms that have been referenced in this chapter: human factors, expert systems, knowledge elicitation, judgment and decision making, and knowledge representation. We anticipate that models of cognition that have been inspired by expertise studies and the NDM paradigm will continue to inform research and development in the field of remote sensing, with applications ranging all the way from training to the design of software decision-support systems and visualization systems. At the same time, research on the cognition of expert remote sensing image interpreters will continue to enrich our models of expert cognition and perception.

References

Ackerman, M., Pipek, V., & Wulf, V. (2003). *Sharing Expertise: Beyond Knowledge Management.* Cambridge, MA: MIT Press.

Argialas, D. P., & Milaresis, G. Ch. (2001). Human factors in the interpretation of physiography by symbolic and numerical representations within an exert system. In R. R. Hoffman & A. B. Markman (Eds.), *The Interpretation of Remote Sensing Imagery: The Human Factor* (pp. 59–82). Boca Raton, FL: Lewis.

Beach, L. R., & Lipshitz, R. (1993). Why classical decision theory is an inappropriate standard for evaluating and aiding most human decision making. In G. Klein, J. Orasanu, R. Calderwood & C. E. Zsambok (Eds.), *Decision Making in Action: Models and Methods* (pp. 21–35). Norwood, NJ: Ablex.

Becerra-Fernandez, I., Gonzalez, A., & Sabherwal, R. (2004). *Knowledge Management: Challenges, Solutions, and Technologies.* Englewood Cliffs, NJ: Prentice Hall.

Burton, A. M., Shadbolt, N. R., Rugg, G., & Hedgecock, A. P. (1990). The efficacy of knowledge elicitation techniques: A comparison across domains and levels of expertise. *Journal of Knowledge Acquisition*, 2, 167–178.

Calderwood, R., Crandall, B., & Klein, G. (1987). Expert and novice fireground command decisions. Report MDA903-85-C-0327, U.S. Army Research Institute, Alexandria, VA.

Crandall, B., Klein, G., & Hoffman, R. R. (2006). *Working Minds: A Practitioner's Guide to Cognitive Task Analysis.* Cambridge, MA: MIT Press.

Department of the Army. (1990). Field Manual for Terrain Analysis (FM-33). Headquarters, Department of the Army, Washington, DC.

Eccles, D., Feltovich, P., & Hoffman, R. (2004). The representation of conceptual knowledge: A model of expert knowledge for terrain analysis. Report on Task 02TA4-SP1-RT4 to the Advanced Decision Architectures Collaborative Technology Alliance. Sponsored by the U.S. Army Research Laboratory, Cooperative Agreement DAAD19-01-2-0009, U.S. Army Research Laboratory, Adelphi, MD.

Ericsson, K. A., Hoffman, R. R., Kozbelt, A., & Williams, M. (2018). *Cambridge Handbook of Expertise and Expert Performance* (2nd ed.). Cambridge, UK: Cambridge University Press.

Gibson, E. J. (1963). Perceptual learning. *Annual Review of Psychology*, 14, 29–56.

Hahn, B. B., Rall, E., & Klinger, D. W. (2002). Cognitive task analysis of the warning forecaster task. Final report for National Weather Service (Office of Climate, Water, and Weather Service, Order No, RA1330-02-SE-0280). Fairborn, OH: Klein Associates, Inc.

Hoffman, R. R. (1984). Methodological preliminaries to the development of an expert system for aerial photo interpretation. Report ETL-0342, Engineer Topographic Laboratories, Ft. Belvoir, VA.

Hoffman, R. R. (1985) What's a hill? An analysis of the meanings of topographic terms. Report. US Army Engineer Topographic Laboratories, Ft. Belvoir, VA.

Hoffman, R. R. (1987). The problem of extracting the knowledge of experts from the perspective of experimental psychology. *The AI Magazine*, 8, 53–67.

Hoffman, R. R. (1990). Remote perceiving: A step toward a unified science of remote sensing. *Geocarto International*, 5, 3–13.

Hoffman, R. R. (Ed.) (1992). *The Psychology of Expertise: Cognitive Research and Empirical AI*. Mahwah, NJ: Erlbaum.

Hoffman, R. R. (2007). The cost of knowledge recovery: A challenge for the application of Concept Mapping. *Presentation at the 51st Annual Meeting of the Human Factors and Ergonomics Society*. Santa Monica, CA: Human Factors and Ergonomics Society.

Hoffman, R. R., Coffey, J. W., Ford, K. M., & Novak, J. D. (2006). A method for eliciting, preserving, and sharing the knowledge of forecasters. *Weather and Forecasting*, 21, 416–428.

Hoffman, R. R., & Conway, J. (1990). Psychological factors in remote sensing: A review of recent research. *Geocarto International*, 4, 3–22.

Hoffman, R. R., Crandall, B., & Klein, G. (2008). Protocols for cognitive task analysis. Report, Florida Institute for Human and Machine Cognition, Pensacola, FL. Retrieved from www.dtic.mil/cgi-bin/GetTRDoc?AD=ADA475456

Hoffman, R. R., & Fiore, S. M. (2007, May/June). Perceptual (re)learning: A leverage point for human-centered computing. *IEEE Intelligent Systems*, pp. 79–83.

Hoffman, R. R., & Hanes, L. F. (2003, July–August). The boiled frog problem. *IEEE Intelligent Systems*, pp. 68–71.

Hoffman, R. R., LaDue, D., Mogil, H. M., Roebber, P., & Trafton, J. G. (2017). *Minding the Weather: How Expert Forecasters Think*. Cambridge, MA: MIT Press.

Hoffman, R. R., Markman, A. B., & Carnahan, W. H. (2001). Angles of regard: Psychology meets technology in the perception and interpretation of nonliteral imagery. In R. R. Hoffman & A. B. Markman (Eds.), *Interpreting Remote Sensing Imagery: Human Factors*. Boca Raton, FL: CRC/Taylor & Francis.

Hoffman, R. R., & Militello, L. G. (2008). *Perspectives on Cognitive Task Analysis: Historical Origins and Modern Communities of Practice*. Boca Raton, FL: CRC/Taylor & Francis.

Hoffman, R. R., & Pike, R. J. (1995). On the specification of the information available for the perception and description of the natural terrain. In P. Hancock, J. Flach, J. Caird & K. Vicente (Eds.), *Local Applications of the Ecological Approach to Human–Machine Systems* (pp. 285–323). Mahwah, NJ: Erlbaum.

Hoffman, R. R., Ward, P., DiBello, L., Feltovich, P. J., Fiore, S. M., & Andrews, D. (2014). *Accelerated Expertise: Training for High Proficiency in a Complex World*. Boca Raton, FL: Taylor & Francis/CRC.

Hoffman, R. R., & Yates, J. F. 2005. Decision(?)making(?). *IEEE: Intelligent Systems, 20*(4), 22–29.

Hutchins, E. (1995). *Cognition in the Wild*. Cambridge, MA: MIT Press.

Jenkins, J. J. (1974). Remember that old theory of memory? Well, forget it! *American Psychologist, 29*, 785–795.

Klein, G. (1989). Recognition-primed decisions. In W. B. Rouse (Ed.), *Advances in Man–Machine Research* (Vol. 5). Greenwich, CT: JAI.

Klein, G., Calderwood, R., & Clinton-Cirocco, A. (1986). Rapid decision making on the fire ground, *Proceedings of the 30th Annual Meeting of the Human Factors Society* (pp. 576–580). Santa Monica, CA: Human Factors Society.

Klein, G., Calderwood, R., & Clinton-Cirocco, A. (2010). Rapid decision making on the fire ground: The original study plus a postscript. *Journal of Cognitive Engineering and Decision Making, 4*, 186–209.

Klein, G., Calderwood, R., & MacGregor, D. (1989). Critical decision method for eliciting knowledge. *IEEE Transactions on Systems, Man, and Cybernetics, 19*, 462–472.

Klein, G., & Hoffman, R. R. (1992). Seeing the invisible: Perceptual-cognitive aspects of expertise. In M. Rabinowitz (Ed.), *Cognitive Science Foundations of Instruction* (pp. 203–226). Mahwah, NJ: Erlbaum.

Klein, G., Moon, B., & Hoffman, R. R. (2006a, November/December). Making sense of sensemaking 2: A macrocognitive model. *IEEE Intelligent Systems*, pp. 88–92.

Klein, G., Moon, B., & Hoffman, R. R. (2006b, July/August). Making sense of sensemaking 1: Alternative perspectives. *IEEE Intelligent Systems*, pp. 22–26.

Klein, G., Phillips, J. K., Rall, E., & Peluso, D. A. (2003). A Data/Frame theory of sensemaking. In [R.] R. Hoffman (Ed.), *Expertise Out of Context* (pp. 113–158). Mahwah, NJ: Erlbaum.

Klein, G., Ross, K. G., Moon, B. M., Klein, D. E., Hoffman, R. R., & Hollnagel, E. (2003, May/June). Macrocognition. *IEEE: Intelligent Systems*, pp. 81–85.

Lowe, R. K. (2001). Components of expertise in the perception and interpretation of meteorological charts. In R. R. Hoffman & A. B. Markman (Eds.), *The Interpretation of Remote Sensing Imagery: The Human Factor* (pp. 185–206). Boca Raton, FL: Lewis.

Miller, G. A. (1986). Dismembering cognition. In S. H. Hulse & B. F. Green Jr. (Eds.), *One Hundred Years of Psychological Research in America* (pp. 277–298). Baltimore, MD: Johns Hopkins University Press.

Mogil, A. M. (2001). The skilled interpretation of weather satellite images: Learning to see patterns and not just cues. In R. R. Hoffman & A. B. Markman (Eds.), *The Interpretation of Remote Sensing Imagery: The Human Factor* (pp. 205–272). Boca Raton, FL: Lewis.

Moon, B. M., Hoffman, R. R., Cañas, A. J., & Novak, J. D. (Eds.) (2011). *Applied Concept Mapping: Capturing, Analyzing and Organizing Knowledge*. Boca Raton, FL: Taylor & Francis.

Newell, A. (1973). You can't play 20 questions with nature and win. In W. G. Chase (Ed.), *Visual Information Processing* (pp. 283–308). New York: Academic.

Novak, J. D. (1998). *Learning, Creating and Using Knowledge*. Mahwah, NJ: Erlbaum.

Schraagen, J. M. C., Chipman, S. F., & Shalin, V. L. (Eds.) (2000). *Cognitive Task Analysis.* Hillsdale, NJ: Erlbaum.

Schraagen, J.-M., Militello, L. G., Ormerod, T., & Lipshitz, R. (Eds.) (2008). *Naturalistic Decision Making and Macrocognition.* London: Ashgate.

Vicente, K. J. (1999). *Cognitive Work Analysis.* Mahwah, NJ: Erlbaum.

Weick, K. E. (1995). *Sensemaking in Organizations.* Thousand Oaks, CA: Sage.

Zakay, D., & Wooler, S. (1984). Time pressure, training, and decision effectiveness. *Ergonomics, 27,* 273–284.

2

Characteristics of Geospatial Photographs in Constructing Human Spatial Knowledge

Pyry Kettunen

CONTENTS

KEYWORDS: *geospatial photograph, spatial knowledge, user experiment, vantage point, visual realism, wayfinding*

2.1 Introduction

Georeferenced photographs have become a ubiquitous resource of spatial knowledge for people now that satellite imagery, aerial orthophotos, oblique aerial photos, and terrestrial photo panoramas are publicly and inexpensively available through the Internet and cover major parts of densely populated regions of the earth. We call these photogrammetric images *geospatial photographs* (abbr. *geophotos*). Geophotos belong to the larger whole of external visuo-spatial representations of large areas in physical reality, which we call *geospatial pictures* (abbr. *geopictures*). People use geophotos for various kinds of spatial activities, such as familiarizing themselves with an environment, as well as for planning and performing spatial actions; for example, planning and navigating routes. Therefore, researching the success of these geospatial photographs in the construction of human spatial knowledge about the visible physical world is imperative, and it is at least as imperative today as it was in 2001, when the previous (first) volume of the current book was published (Hoffman & Markman, 2001).

In this manuscript, the term *geospatial photograph* refers to a photograph (including satellite images) or a composition of photographs representing any portion of the earth's surface. This chapter discusses the characteristics of geophotos in relation to the human acquisition of spatial knowledge by reviewing theoretical and experimental knowledge in the literature on human spatial cognition. To accomplish this, we introduce and use our previously developed literature-based framework for assessing the support of geopictures for spatial knowledge acquisition. A review of literature is presented with a short history of spatial knowledge research (Section 2.2), followed by the introduction of the assessment framework (Section 2.3). The framework is then applied to three current types of geophotos for assessing the effects of vantage point, number of visible vertical features, and visual realism in these photos for supporting landmark, route, and configurational types of spatial knowledge (Section 2.4).

Different types of geophotos are differently suitable for different kinds of spatial activities, because they allow for divergent profiles of acquired spatial knowledge. Therefore, this chapter also discusses the suitability of three different geophoto types for use in three common human geospatial activities and proposes particularly suitable types of photos for each activity (Section 2.5).

The number of published experimental user studies on many kinds of geophotos appears surprisingly low. The chapter identifies the gaps in the previous research in addressing the user studies of geophotos from a spatial cognition perspective and proposes research challenges for possible future work for completing the gaps in experimental research (Section 2.6). Finally, a concluding summary is given about the contents of the chapter (Section 2.7).

2.2 Human Spatial Knowledge

2.2.1 Brief History of Spatial Knowledge Research

Spatial knowledge research has evolved in parallel with the development of *spatial cognition* research, which emerged following the birth of psychology as a discipline in the nineteenth century. In accordance with psychology, initial research on spatial cognition leaned on so-called armchair experiments, in which psychologists reasoned by pure deduction how people think about space around them. For example, Trowbridge (1913) discussed why people maintain faulty representations of spatial reality, or "imaginary maps," in their minds. Empirical research on spatial cognition began with rat experiments in the 1920s, which culminated with the introduction of the concept of *cognitive maps* by Tolman (1948). Tolman proved that rats are able to create shortcuts that they have not learned by direct experience, which necessitates them forming an abstract conception of space. He reasoned that people do the same, and decades of intensive experimental cognitive mapping research followed based on this conceptualization.

Downs and Stea (1977) reviewed the collective knowledge concerning cognitive mapping and defined cognitive maps comprehensively as "a process composed of a series of psychological transformations by which an individual acquires, codes, stores, recalls, and decodes information about the relative locations and attributes of phenomena in his everyday spatial environment" (p. 8). The maturity of the discipline led to the description of three types of knowledge: *landmark, route*, and *configuration knowledge* (Siegel & White, 1975; see next section). These concepts are still valid and used in spatial cognition research, although they have been developed further during the latest decades.

To complement traditional behavioral experimental methods, neurocognitive studies based on brain imaging have become an important source of evidence for spatial cognition research since the 1970s. Such techniques as computer axial tomography (CAT), positron emission tomography (PET), and functional magnetic resonance imaging (fMRI) have made it possible to measure the activation of different cerebral lobes during spatial thinking and to prove how spatial knowledge builds in human memory (e.g., Janzen & van Turennout, 2004; O'Keefe & Nadel, 1978). Neural understanding of spatial cognition has grown quickly in the twenty-first century, as the existence of spatial neurons as well as the roles of cerebral lobes have been discovered and verified (e.g., Sargolini et al., 2006).

2.2.2 Structure of Spatial Knowledge

According to a well-known model of human spatial knowledge by Siegel and White (1975), people conceive environmental-scale spaces at three levels: landmark, route, and configuration knowledge. *Environmental-scale spaces* are considered to be wider than would be observable from one point of view but still apprehensible through direct experience (Montello, 1993).

Landmark knowledge is composed of information about individual physical landmark features, that is, "unique configurations of perceptual events" (Siegel & White, 1975; p. 23). As human "perceptual events" of the physical world are centrally visual, landmark knowledge is mainly stored in visual memory. People recognize places using landmark knowledge, the first type of spatial knowledge that develops in children (Siegel & White, 1975). When people are exposed to an unfamiliar environment, they begin to build landmark knowledge from the very beginning. In Siegel and White's spatial knowledge model, landmark knowledge establishes the base of all spatial knowledge, and spatial learning begins with it. This constitutional role of landmark recognition in human spatial cognition is supported by wide experimental evidence (e.g., Janzen & van Turennout, 2004; Passini, 1984; Richardson, Montello, & Hegarty, 1999). Passini even showed *in situ* that people with insufficient configuration knowledge (see the paragraph after next) of a shopping mall were able to find their way using only passive landmark knowledge that they could not actively recall.

Route knowledge consists of linearly ordered information about paths and surrounding landmarks, that is, "sensorimotor routines for which one has expectations about landmarks and other decision points" (Siegel & White, 1975, p. 24). In an unfamiliar environment, route knowledge is built by forming a sequence of observed landmarks along a traveled route. Route knowledge allows people to find their way from place to place along single pathways.

Configuration knowledge is spatially structured information of routes and landmarks "that gives its owner an advantage in way-finding and organizing experience" (Siegel & White, 1975, p. 24). In configuration knowledge, observed landmarks and routes are combined into new ones and interconnected, so that comprehending environmental patterns and planning untraveled routes becomes possible. The acquisition of configuration knowledge in an unfamiliar environment takes considerably longer than the acquisition of landmark and route knowledge: sufficient amounts of these less sophisticated types of knowledge are needed before configuration knowledge can take shape. The concept of configuration knowledge is similar to concepts termed *survey knowledge* and *cognitive map*.

To summarize, in the widely adopted Siegel and White (1975) model, landmark, route, and configuration knowledge form the whole of human spatial knowledge, with landmark knowledge as the foundation for the other two types.

2.2.3 Measuring Spatial Knowledge

Measuring the extent and accuracy of human spatial knowledge usually focuses on route and configuration knowledge, as these two provide the largest benefits for concrete human spatial activity and enable creative spatial thinking. So far, measuring appears to have relied on indirect behavioral methods, since we are still far from being able to read the detailed contents of human memory directly. Indirect measures of spatial knowledge include *performance measures*, such as traversal time, turning error count, and distance and direction estimates, as well as *descriptive measures*, such as sketch maps and verbal route descriptions (see Newcombe, 1985). Traversal time and turning error counts are particularly suited for measuring route knowledge, whereas distance and direction estimates measure configuration knowledge (e.g., Richardson et al., 1999; Thorndyke & Hayes-Roth, 1982). In turn, descriptive measures are used to assess both route and configuration knowledge, because they are collections of direct conceptualizations of space by the subjects themselves (e.g., Ishikawa & Montello, 2006; Kettunen, 2014).

The problem of indirect measurement of spatial knowledge is that different measures easily produce seemingly conflicting results due to the difficulty of controlling latent independent variables, such as the particular structure of the experimental space, environmental conditions, personal characteristics of subjects, or varying interpretations of the task by subjects. Nonetheless, the still-growing repertoire of empirical studies on spatial knowledge has fostered the capability to reliably measure individual spatial knowledge. For example, concurrent think-aloud (Boren & Ramey, 2000) and sketch map drawing (Newcombe, 1985) have been shown to be reliable sources of information on people's spatial knowledge despite—or because of—their descriptive and subjective nature, although suspicions have been raised regarding these methods (e.g., Smagorinsky, 1998).

Due to the development of three-dimensional (3D) computer graphics since the 1990s, virtual environments have become a popular platform for measuring spatial knowledge in controlled interactive settings that resemble the real world, and the development is continuing with the progress of virtual headsets. Real-world measures can be applied also in virtual environment experiments (e.g., Waller, Hunt, & Knapp, 1998), although virtual environments cannot currently reproduce the real world sufficiently to replace *in situ* experiments completely (van der Ham, Faber, Venselaar, van Kreveld, & Löffler, 2015)—the full assemblage of attractors of attention in reality is too multifaceted to model (see Keuth, 1976).

2.2.4 Spatial Knowledge Acquisition

The types of spatial knowledge proposed by Siegel and White (1975) are still considered valid today, but models of the knowledge acquisition have changed. The idea of consecutive accumulation of each knowledge type has been replaced by the continuous accumulation of each type, beginning from the first exposure to an unfamiliar environment (Ishikawa & Montello, 2006). This *continuous model* of spatial knowledge acquisition additionally takes into account human understanding of the usual structures of space and knowledge from external geospatial representations, such as verbal route descriptions, maps, or aerial photographs.

Landmark knowledge acquisition has been an intensively studied topic, because the other knowledge types rely on it (see Richter & Winter, 2014). While the term *landmark* has been associated with particularly salient features in the environment since the 1960s (Lynch, 1960), research on spatial cognition has also widely adopted a definition of *landmark* as any physical feature of environment that is used in spatial thinking (e.g., Brosset, Claramunt, & Saux, 2008; Denis, 1997). In this chapter, we refer to a landmark according to the latter definition.

What features are used as landmarks depends, importantly, on the type of environment, surrounding conditions, locomotion modality, and the user's personal experience. In urban environments, distinguishable constructions, such as particular buildings, plantings, or crossings (e.g., Denis, 1997), are used as landmarks, whereas in nature, any construction or pathway easily becomes a landmark (see Kettunen, 2014), and in wilderness, only particular vegetations or landforms are used as landmarks. Diurnal patterns also affect what types of landmarks are used; for example, illuminations are used more heavily at night than during the daytime (see Kettunen, 2014). Landmarks for car driving are different from pedestrian landmarks, and expertise on some specific environmental features, such as vegetation, may allow a person to make useful landmarks of features of that type. Landmarks can also be categorized into global and local landmarks, based on the distance over which they can be used, or into visual, cognitive, and structural landmarks, based on their salient factors (see Richter & Winter, 2014).

In addition to empirical landmark studies, numerous studies have been conducted to investigate the acquisition of spatial knowledge from direct experience in varied environments and conditions by multiple types of subjects and tasks. Most of these studies have been organized in wayfinding settings, because wayfinding is a cognitively challenging spatial task that strongly depends on spatial knowledge and a task that people know well due to daily experience. For a full review of these studies, see Kettunen (2014).

Most of the experiments on the acquisition of spatial knowledge from direct experience have been conducted in urban environments, and many of these are according to the classification of propositions in route descriptions by Denis (1997). Denis modeled a formula of good route descriptions based on his collection of skeletal descriptions from the subjects of the study and showed that landmarks played the most central role in these descriptions. Applications of Denis's method include, for instance, the study of Rehrl, Leitinger, Gartner, and Ortag (2009), who collected route descriptions *in situ* in an unfamiliar environment for the subjects, so that the wayfinding situation maximally resembled a real use case of a navigator. The role of landmarks was emphasized, and almost half of the mentions of landmarks were connected to action directives for proceeding on the route.

Nature environments are challenging both for spatial cognition and for experimentation with it, so only a couple of studies have experimented with the acquisition of spatial knowledge in nature. Brosset et al. (2008) investigated route descriptions of orienteers after an orienteering race using the method of Denis (1997) and found that even higher frequencies of landmarks were used in nature than in the campus environment used by Denis (1997) in his experiments, and that these were often linear features of the terrain. In a pair of studies, Kettunen and colleagues studied route descriptions and examined the effect of summer, winter, day, and night (for summary, see Kettunen, 2014), finding even higher frequencies of landmark use than Brosset et al. Kettunen and colleagues also found that structures, passages, and water features were the most used landmarks over all conditions and that used landmark types varied between conditions *in situ* but not in recall, suggesting that people build landmark knowledge independently of the conditions.

Geospatial pictures, such as maps and aerial photographs, are efficient means to acquire spatial knowledge, particularly route and configuration knowledge, as they provide views of large environments in condensed and readily interpreted forms. However, cognitive experiments on geopicture reading have a shorter tradition and a smaller body of knowledge than direct-experience experiments of spatial cognition. One of the early experiments was the study of Thorndyke and Hayes-Roth (1982), which showed, indoors, that maps allowed accurate configuration knowledge, whereas direct experience allowed more robust route knowledge. Richardson et al. (1999) replicated the study of Thorndyke and Hayes-Roth with the additions of a virtual 3D environment and control for map alignment. The results on route and configuration knowledge were replicated, and map alignment was found to be a key factor in map reading. Virtual 3D environments were found to be weak sources of spatial knowledge, which other studies have replicated (e.g., Oulasvirta, Estlander, & Nurminen, 2009; Waller et al., 1998). Waller et al. (1998) discovered in a maze environment that maps allow the most rapid acquisition of accurate route and configuration knowledge, but the repetition of a route in direct experience ends up with similarly accurate knowledge. The superiority of maps for configuration knowledge and their inferiority for route knowledge were also found with real navigation tasks in the study of Münzer, Zimmer, and Baus (2012).

Geophotos have been investigated for the acquisition of spatial knowledge in only a few studies. Beeharee and Steed (2006) showed the benefits of street photos for route instructions in a real navigation study in which participants used photos for reassurance at decision points of the route, whereas they preferred a two-dimensional (2D) map for an overall view of the route. Hile et al. (2008), as well as Liu et al. (2009), also found participants checking landmarks from street photos at decision points. Hile et al. (2008) additionally showed that non-augmented aerial orthophotos were more helpful for route planning and navigation than ground-level photos. However, none of these studies provided sufficient statistical methodology or number of participants to draw comprehensive conclusions about the amount and accuracy of acquired spatial knowledge, and there is still a lack of evidence about the acquisition of spatial knowledge from geophotos. In the forthcoming sections of this chapter, we elaborate the potential of geophotos for communicating spatial knowledge based on the existing geopicture literature.

2.3 A Framework for Evaluating Spatial Knowledge Acquisition from Geopictures

While we are missing comprehensive empirical evidence about the usefulness of geophotos for acquiring spatial knowledge, it is possible to identify general factors that affect the acquisition. Based on literature, we previously developed a framework of significant *pictorial parameters* that can explain how useful geopictures are in supporting the acquisition of different types of spatial knowledge (see Kettunen, 2014). In this section, we introduce the parameters *vantage point, number of visible vertical features,* and *visual realism,* together with experiments in previous studies that show the impacts of these parameters on the human acquisition of spatial knowledge.

2.3.1 Vantage Point and Number of Visible Vertical Features

In our framework, the *vantage point* of a geospatial picture is the vertical viewpoint or perspective of the picture on the terrain, defined by the vertical viewing angle of the picture to the ground. We categorize vantage points of geopictures into three classes: aerial vertical, aerial oblique, and ground horizontal vantage points (Figure 2.1). In turn, the *number of visible vertical features* is the number of features perpendicular to the ground that are visually represented on the picture. The number of visible vertical features may vary continuously from non-existent to comprehensive, the latter meaning the case in which all possible vertical features are depicted in the picture.

It is worth noting that a geopicture may have a single vantage point for the whole picture or a composition of multiple vantage points for different features to emphasize some feature types over others. For instance, attractions are highlighted on a typical tourist map using an aerial oblique vantage point, while the base map is depicted from an aerial vertical vantage point. The number of visible vertical features increases while the vantage point gets lower.

The number of visible vertical features is dependent on the vantage point of a geopicture, as vertical surfaces are not represented on a picture with an aerial vertical vantage point. In contrast, there may be a large number of vertical surfaces in a picture with a ground horizontal vantage point (see Figure 2.2). The number of visible vertical features is also dependent on the content or data that is depicted: when representing 2D data, the number of visible vertical features is non-existent independently of the vantage point, whereas, for example, for a tourist map, as described earlier, 3D buildings increase the number of vertical features. On the other hand, for a terrestrial

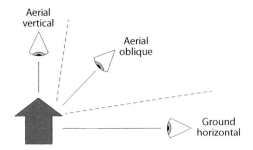

FIGURE 2.1
Three classes of vantage points (vertical viewpoints or perspectives).

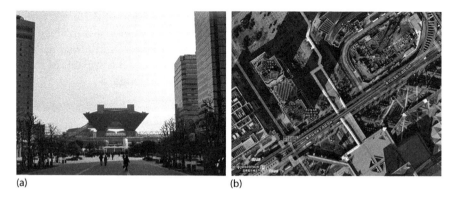

(a)   (b)

FIGURE 2.2
Number of visible vertical features varies considerably between (a) ground horizontal (b) and aerial vertical (same area: © Google, Digital Earth Technology, DigitalGlobe, GeoEye, The Geoinformation Group, and ZENRIN) vantage points.

photo, the number of visible vertical features is the same as in the full 3D real world.

Experimental studies of geopictures have shown a strong impact of the ensemble of the vantage point (whether it is vertical, oblique, or horizontal) and the number of visible vertical features on the types and amounts of spatial knowledge that people acquire. Particularly, an aerial oblique vantage point with a large number of visible vertical features has been shown to successfully support the acquisition of the three types of spatial knowledge due to the presence of each type on the same picture (Fontaine, 2001; Plester, Richards, Blades, & Spencer, 2002): the recognition of landmarks is possible on visible vertical surfaces, and landmarks can be easily associated with the spatial structures of routes and the configuration of the environment due to the wide spatial extent enabled by the view from above.

The aerial vertical vantage point also allows successful acquisition of spatial knowledge according to empirical studies. Particularly, vertical view geopictures appear to allow configuration knowledge to be successfully acquired due to the presence of a spatial overview and some landmark knowledge. This combination seems to enable the necessary spatial knowledge to be acquired for success in wayfinding (e.g., Hile et al., 2008; Oulasvirta et al., 2009; Waller et al., 1998). People have been found to use the ground horizontal vantage point mostly in decision points of routes where detailed landmark knowledge is beneficial (e.g., Hile et al., 2008; Oulasvirta et al., 2009). These studies also show that the ground horizontal vantage point is insufficient for providing linking information between landmarks.

2.3.2 Visual Realism

Visual realism is the optical resemblance of a geopicture to the physical world, which is inverse to the degree of abstraction in the picture (see Figure 2.3).

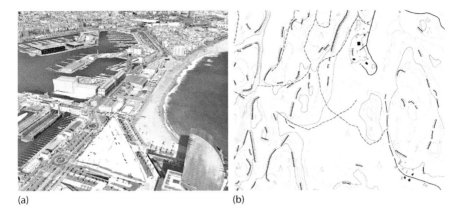

(a)   (b)

FIGURE 2.3

High visual realism in (a) a photorealistic rendering (© Google and Institut Cartogràfic de Catalunya) and (b) a lower visual realism in a topographic map (© National Land Survey of Finland).

Visual realism varies between photorealistic and abstract extremes, for example, from orthophotos to topographic maps (Hoarau, 2015). Different levels of visual realism may be used in a single geopicture (sometimes referred to as a *hybrid* picture), for instance, to highlight significant features with more or less abstract depiction than in the other parts of the picture.

Visual realism has been actively studied in the context of virtual reality environments, where it has been shown to have a positive effect on the acquisition of spatial knowledge, to increase the sense of presence, and to be preferred by users (e.g., Lokka & Çöltekin, 2017; Meijer, Geudeke, & van den Broek, 2009). However, high visual realism also reduces the efficiency of geopicture interpretation and even causes negative effects on the acquisition of accurate spatial knowledge due to superfluous spatial information on the picture (e.g., Hegarty, Smallman, & Stull, 2012; Lokka & Çöltekin, 2017; Meijer et al., 2009). Therefore, the benefits of high visual realism must be carefully considered based on the use case of a geopicture.

The effect of visual realism in geophotos on the acquisition of spatial knowledge has been studied only marginally. Haynes et al. (2007) compared aerial oblique geophotos with visually unrealistic 2D and 3D maps in a task about volcanic hazard evaluation on an island. They found that geophotos provided local people with more convenient interpretability than maps due to the self-explanatory representation of the photos and provided people with more accurate configuration knowledge. The result is similar to another study with naive subjects, namely children, in which children were able to use photographs more successfully than drawn maps for search tasks on the pictures (Plester et al., 2002). In a recent study, Çöltekin, Francelet, Richter, Thoresen, and Fabrikant (2017) found that routes on abstract maps are remembered better than on photorealistic satellite images, particularly by spatially talented participants, which implies support for acquisition of spatial knowledge by abstraction. It must be noted that thus far, most of the geophoto studies have investigated visual realism in contrast to non-realism, and comprehensive quantitative evidence of visual realism in photographs as an enabler of spatial knowledge is still missing.

2.4 Geophotos in Spatial Knowledge Acquisition

Geophotos are the most realistic form of geopictures, as they clone human visual experience onto external media. However, the use of geophotos as a source of spatial knowledge considerably differs from direct experience due to photos being static representations of visual reality with invariant time and location. Therefore, the assessment of geophotos in the context of spatial knowledge acquisition is important for understanding how people can successfully acquire knowledge from geophotos and consequently, how geophotos can be made more supportive of spatial knowledge acquisition. This section elaborates the topic both in general and separately for the three types of geophotos that are commonly seen in geospatial applications, such as web maps or navigators. Our geophoto assessment takes the perspective of people's acquisition of knowledge in everyday situations on environmental scales, excluding satellite images on geographic scales. The sparse existing empirical literature and our previously developed assessment framework (see Kettunen, 2014) form the basis for the assessment results, the summary of which is illustrated in Figure 2.4.

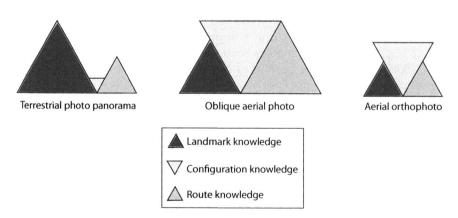

FIGURE 2.4
Potential for the acquisition of the types of spatial knowledge from the three types of geophotos. Sizes of the triangles qualitatively illustrate the assessed support by the proposed assessment framework.

2.4.1 General Characteristics of Geophotos for Acquiring Spatial Knowledge

With regard to the pictorial parameters of our framework, the common parameter of all geophotos is the level of visual realism (i.e., they are all photorealistic), whereas the vantage point and number of visible vertical features may range over their scales. However, for geophotos, the number of visible vertical features is directly related to the vantage point: the more horizontal the vantage point, the higher the number of visible vertical features. Accordingly, three types of geophotos can be named, based on the vantage points that correlate with these in our framework: photos taken from the aerial vertical, oblique aerial, and ground horizontal vantage points (as shown in Figure 2.1).

Photorealism makes the features in geophotos readily recognizable by people due to their resemblance to our continuous visual perception of real-world features. Even with less common vantage points in everyday life, such as the aerial vertical one, the colors and forms of the features in the geophoto match the real-world features and are thus easy to interpret, leading to realistic internal representations of the environment, even for young children (Plester et al., 2002) or cognitively impaired people (Liu et al., 2009). However, decreasing spatial resolution can complicate interpretation, as features merge together in single pixels and become depicted in geometric forms or colors that are unfamiliar to everyday visual experience. Consequently, low-resolution geophotos, particularly from aerial vantage points, require the interpreter to have prior knowledge, so that he or she can recognize these features and connect them from a geophoto to real-world objects. In terms of spatial knowledge, this recognition knowledge allows the acquisition of landmark knowledge on geophotos and functions as the enabler of other forms of spatial knowledge.

The fact that the represented features are similar to their real-world counterparts can be differently helpful to people in the reading of geophotos based on their expertise and previous knowledge. In some cases, users can extract features relevant to their specialized knowledge. For example, a user erudite in plant biology may be able to recognize plants as landmarks, although other users might have difficulties in distinguishing between plants visible on a geophoto.

Visual resemblance to reality (or verisimilitude; see Keuth, 1976) also places some restrictions on geophotos as compared with other kinds of geopictures, such as maps. The most obvious and influential restriction is the impossibility of representing feature types coherently throughout a photo. This may be caused by the occlusion of features behind other, possibly irrelevant, features, such as clouds or foliage, but also by extreme lighting, such as bright reflections or shadows, or by variation of resolution on a photo, which prevents the depiction of small features on some parts of the photo. These shortcomings of information coherence directly reflect the completeness of acquirable spatial knowledge.

Another restriction introduced by visual realism, or the visual resemblance to the real world, unless augmented, on geophotos is the lack of any thematic information. The lack of thematic information forces the viewers themselves to search for categorical, statistical, or other knowledge to understand the spatial structure of the picture, particularly in relation to man-made features. For example, it may be impossible to recognize categories of transport networks or buildings, or the depth of a river to be crossed, which can make it difficult to build detailed and reliable route knowledge based on the geophoto.

Temporal invariance is also a limitation of geophotos for building widely useful spatial knowledge, as time-dependent natural changes of the environment are not necessarily pictured. Temporal changes make environments visually very different depending on the time of taking a photo for such reasons as illumination or absence of vegetation in the winter. The changes cause significant variation in the types of landmarks that can be used (see Kettunen, 2014). Mental transformation of spatial knowledge from the original photo to the time of use can be challenging and may result in difficulties in applying that knowledge.

2.4.2 Terrestrial Photo Panoramas

By *terrestrial photo panoramas*, we mean collages of wide panoramic photos with a ground horizontal vantage point (first-person perspective) that depict the environment approximately at eye level (see Figure 2.5). Terrestrial photography has been used in geography-related domains for a long time as a source of information. Modern street photos are typically captured in panoramic fashion for a larger coverage. We categorize terrestrial photo panoramas as geophotos, since they represent complete wayfinding environments, for example, in web map applications such as Google Street View or Bing Streetside. Such photo collages are typically presented through an interactive

FIGURE 2.5
Example of an extract from a terrestrial photo panorama (© Google).

user interface that continuously presents composite images from single wide-angle photos to create a seamless user experience. This type of geophoto is most intuitive for people to read, as it directly clones our everyday visual flux.

2.4.2.1 Landmark Knowledge

Terrestrial photo panoramas allow the maximal acquisition of landmark knowledge due to their ground horizontal vantage point, which pictures a comprehensive number of visible vertical features in high ground resolution. Characteristic details of landmarks are depicted, which makes landmarks easy to recognize and recall. Experimental studies on the use of street photos for real-world navigation support this assessment by our framework, as these studies have found the greatest use of street photos for checking details of landmarks, for instance, at the decision points of routes (Beeharee and Steed, 2006; Hile et al., 2008; Liu et al., 2009).

2.4.2.2 Route Knowledge

Terrestrial photo panoramas may provide useful route knowledge if the extent of the vista in a photo is wide enough and, particularly, if photos are used in a compositing interactive user interface. However, a relatively low number of landmarks are visible at one time, and the acquisition of route knowledge depends heavily on the spatial skills of the user, who needs to memorize the locations of landmarks and their interconnections to build useful route knowledge.

2.4.2.3 Configuration Knowledge

Terrestrial photo panoramas are likely to minimally support the acquisition of configuration knowledge, because they depict only one vista at a time. The conception of the structure of the environment depends on the spatial skills of the user, as perceived landmarks and their interconnections must be kept in the memory, based on which further interconnections must be reasoned. Building broad configuration knowledge based on terrestrial photo panoramas also requires considerable amounts of time and effort for exploring large numbers of photos.

2.4.2.4 Assessment Summary

Terrestrial photo panoramas contribute primarily to the acquisition of detailed landmark knowledge of the environment. They can be used for building useful route knowledge with some effort, but acquiring configuration knowledge is overly resource intensive compared with other types of geophotos. Experiments conducted until now support this assessment by our framework, as described earlier, but experiments on route and configuration knowledge are yet to be carried out with terrestrial photo panoramas.

2.4.3 Oblique Aerial Photos

An *oblique aerial photo* is a photo from an aerial oblique vantage point, typically taken from low altitude with high ground resolution (see Figure 2.6). Oblique aerial photos are widely available in web map applications today, in which they are often provided in mosaics and in multiple horizontal viewing directions with an interactive user interface. Oblique aerial photos resemble common human experience vistas from viewing places at high altitudes and from aircraft and are thus presumably easy for people to interpret.

2.4.3.1 Landmark Knowledge

Oblique aerial photos carry large amounts of useful landmark information due to their oblique vantage point, which allows a high number of visible vertical features. Their potential support for the acquisition of landmark knowledge is thus high, although resolution may restrict the amount of visible details on landmarks, and tall objects may occlude important landmark features on the ground level. These restrictions, however, can be overcome to some extent by using an interactive browsing interface with photos from multiple directions. There is some experimental evidence supporting our position by the assessment framework; that is, people can successfully acquire landmark knowledge from oblique aerial photos (Plester et al., 2002).

2.4.3.2 Route Knowledge

Oblique aerial photos have the potential to optimally support the acquisition of route knowledge, because they contain a high amount of landmark information and allow instant viewing of spatial interconnections between the landmarks. However, our literature search did not find any explicit experiments on route navigation using oblique aerial photos.

FIGURE 2.6
Example of an oblique aerial photo (© Microsoft).

2.4.3.3 Configuration Knowledge

Oblique aerial photos can also provide optimal support for acquiring configuration knowledge, as they simultaneously depict landmark and route information in the configurational spatial structure of the environment, as represented by the viewpoint from above. This position (from the assessment by our framework) is supported by a previous study, in which Haynes, Barclay, and Pidgeon (2007) found that people readily interpret oblique aerial photos in tasks on geographical search and map skills, which indicates efficient acquisition of configuration knowledge.

2.4.3.4 Assessment Summary

Oblique aerial photos are potentially the best type of geophotos for spatial knowledge acquisition, because they allow both maximal route and configuration knowledge from a single photo, as well as only slightly restricted landmark knowledge, which can be compensated by interactive browsing software. The few experiments on oblique aerial photos support this position by our framework, but complementary empirical evidence should be collected, particularly for acquiring route knowledge.

2.4.4 Aerial Orthophotos

An *aerial orthophoto* is a rectified mosaic of single aerial photos orthogonal to the ground surface (see Figure 2.7). Rectification is done according to a digital surface model, so that scale on the photo becomes invariant. Only regions close to the nadir of the photo are used for the orthophoto mosaic to maintain the orthogonal viewing angle, and so, dozens of photos may be required to build a large orthophoto. Compared with human visual experience of the real world, orthophotos resemble vertical viewing of ground

FIGURE 2.7
Example of an aerial orthophoto (© National Land Survey of Finland).

from the air, for example, from an airplane, which makes the visual form familiar to many people, although certainly not as many as the aerial oblique and ground horizontal vantage points.

2.4.4.1 Landmark Knowledge

Aerial orthophotos may allow the acquisition of some landmark knowledge, particularly if their resolution is high enough for feature recognition, but this is not always the case. The aerial vertical vantage point makes landmark recognition more challenging than other types of geophotos, because the number of visible vertical features is very small, and landmarks are depicted from a direction that is unusual in human experience. Experimental studies have found that the other types of geophotos are preferred by people for acquiring landmark knowledge (e.g., Hile et al., 2008), but the effect on performance is still unstudied.

2.4.4.2 Route Knowledge

Aerial orthophotos may allow the acquisition of some route knowledge, because interconnections between landmarks can be viewed if appropriate features are visible. However, transport networks, as well as landmarks, may often be occluded due to vegetation or buildings, which potentially reduces the inclusiveness of the acquired knowledge. In open areas, useful route knowledge can be acquired, which has been demonstrated empirically by Hile et al. (2008).

2.4.4.3 Configuration Knowledge

Aerial orthophotos have considerable potential for providing maximal configuration knowledge because of their comprehensive and equal depiction of all terrain due to the aerial vertical vantage point. In the case that landmark and route information is sufficiently visible, this vantage point ensures the perceptibility of interconnections between landmark and route features, potentially resulting in comprehensive configuration knowledge. Seemingly, no experiments have explicitly considered the acquisition of configuration knowledge from aerial orthophotos at this time.

2.4.4.4 Assessment Summary

There are reasons to believe that aerial orthophotos would best support the acquisition of configuration knowledge. They should also enable the acquisition of landmark and route knowledge to some extent, depending on the resolution and occlusion of features on a photo. Experimental studies on the acquisition of spatial knowledge from aerial orthophotos are too few to give empirical evidence about these propositions by our framework.

2.5 Geophotos in Common Geospatial Activities

Geophotos have benefited most kinds of human geospatial activities since the invention of photography in the early nineteenth century, but they have become a ubiquitous everyday tool only since the birth of the Internet and web map applications in the 1990s; available to everybody with an Internet connection. Nowadays, Internet users can use geophotos from almost everywhere in quantities and qualities surpassing what was available only to professionals two decades ago. Therefore, it is important to reflect on the feasibility of different kinds of geophotos for different kinds of spatial activities so as to be able to use an optimal photo for each activity and to develop forms of geospatial photography toward even better utility. We present here three common spatial activities that almost everybody carries out sometimes—wayfinding, free environmental exploration, and environmental functions planning—and their requirements for the three types of spatial knowledge (Table 2.1). Following each, we briefly assess the utility of the three types of geophotos for these activities.

2.5.1 Wayfinding

Wayfinding is a very common human spatial activity on the environmental scale. It means determining and following routes between origins and destinations (Golledge, 1999). Wayfinding contains on-route spatial planning and decision making, and these make it the intellectual component of navigation (whereas locomotion is the physical component) (Montello, 2005). Wayfinding requires place recognition for identifying known places and for enabling spatial updating (together with dead reckoning), which enables awareness of one's location in the environment (see Montello, 2005). Consequently, landmark and route knowledge are necessary for wayfinding, whereas configuration knowledge will give additional benefit for a wayfinder (see Table 2.1).

TABLE 2.1

The Three Considered Common Geospatial Tasks and Their Requirements for the Three Types of Spatial Knowledge

	Landmark Knowledge	Route Knowledge	Configuration Knowledge
Wayfinding	Necessary	Necessary	Beneficial
Free environmental exploration	Beneficial	Beneficial	Beneficial
Environmental functions planning	Beneficial	Necessary	Necessary

2.5.1.1 Terrestrial Photo Panoramas

As terrestrial photo panoramas allow accurate landmark knowledge and can potentially provide good route knowledge, they can be very useful as wayfinding aids, especially for route following at decision points, which is shown by empirical navigation studies (Beeharee & Steed, 2006; Hile et al., 2008; Liu et al., 2009). However, their use for route planning can be challenging, because the acquisition of beneficial configuration knowledge requires considerable learning effort with multitudes of photos, even if convenient browsing software can be used.

2.5.1.2 Oblique Aerial Photo

According to our earlier assessments of this geophoto type and a couple of experimental findings (Fontaine, 2001; Plester et al., 2002), oblique aerial photos should be very useful for wayfinding due to their efficient simultaneous coverage of the three types of spatial knowledge. Landmarks can be easily learned and recognized to support wayfinding, and route and configuration knowledge can be efficiently built from the same elevated viewpoint. The usefulness of oblique aerial photos can be enhanced by interactive user interfaces that allow multiple horizontal viewing directions to depict areas occluded behind tall features from some directions.

2.5.1.3 Aerial Orthophoto

Aerial orthophotos should particularly support route determination for wayfinding activities, as they depict terrain homogeneously from the vertical viewpoint and thus provide the wayfinder with good configuration knowledge. However, due to the low amount of landmark information in the aerial orthophotos (which is fundamental for acquiring configurational and route knowledge), wayfinders may find it difficult to recognize landmarks during navigation. Indeed, experimental evidence shows that people prefer aerial orthophotos over street photos for route planning but easily dismiss them for landmark recognition (Hile et al., 2008). However, performance studies on the topic are missing.

2.5.1.4 Suitability Summary

Oblique aerial photos are clearly the most promising geophoto type for wayfinding due to their wide coverage of spatial information in relation to types of spatial knowledge, according to both our assessment based on the proposed framework and empirical evidence cited in previous paragraphs. Aerial orthophotos and oblique aerial photos can each successfully serve for one sub-activity of wayfinding, route determination and landmark recognition, respectively, but each is lacking properties to sufficiently support the other.

2.5.2 Free Environment Exploration

We call the activity of studying one's environment in a random and unrestricted way to familiarize oneself with the environment *free environmental exploration*, which is a common human spatial activity practiced when one wishes to learn about an unfamiliar area. Exploration may be performed through direct experience in the environment, but today, it begins most typically using geopictures, such as geophotos or maps. In free environmental exploration, the attention is guided by visual attractors of the environment and by personal interests, which together define what features the explorers will remember as landmarks and what routes and areas will attain particular roles in the explorers' configuration knowledge. No type of spatial knowledge is necessary for free environmental exploration, but all types are beneficial for further exploration (see Table 2.1).

2.5.2.1 Terrestrial Photo Panoramas

According to our previous geophoto assessment within the proposed framework and suggestions from the literature (e.g., Beeharee and Steed, 2006), terrestrial photo panoramas fit well for experience-oriented free environmental exploration, as they provide visual experience similar to everyday human visual reality. However, exploring large areas is laborious even with good browsing software, and there is a possibility that the acquired landmark knowledge will develop into an incoherent mental representation of visual cues about landmarks and experiences of places without structured route or configuration knowledge. For example, this may occur if the explorer is unconsciously following attention-guiding cues without any focus on remembering spatial structures later.

2.5.2.2 Oblique Aerial Photo

Oblique aerial photos, again, provide the most promising option for free environmental exploration, as they not only enable accurate and comprehensive landmark, route, and configuration information but also provide a visual experience that easily relates to that of real life (e.g., Plester et al., 2002). Occlusion behind tall features may challenge the explorer, but interactive user interfaces with multiple horizontal viewing directions can enhance the effectiveness of exploration. The downside of such browsing interfaces is that they demand some cognitive resources for interaction and for spatial rotations that occur when changing the viewing direction. Still, we believe that even a sole reading of a single oblique aerial photo can familiarize the explorer very well with a previously unknown environment.

2.5.2.3 Aerial Orthophotos

As useful providers of configuration knowledge, aerial orthophotos maintain a high potential for supporting free environmental exploration but may require some interpretation skills from the explorer for recognizing features on the orthophoto to be used as landmarks, particularly on low-resolution photos. Browsing software with resolution scaling may aid in landmark recognition. Some occlusion of features cannot be avoided in aerial orthophotos, which makes the acquired spatial knowledge partially incomplete.

2.5.2.4 Suitability Summary

Our qualitative assessment based on the proposed framework, as well as guidelines from literature cited in the preceding paragraphs, suggests that oblique aerial photos are the most promising among the examined geophoto types for free environmental exploration also, because they cover the three types of spatial knowledge. Nonetheless, in this case, aerial orthophotos also have a high potential for successful exploration due to their coherence in scale, low amount of occlusion, and large areas depicted by a single image, although recognition of landmarks may be restricted. Terrestrial photo panoramas appear least suited for free environmental exploration because of the large amount of effort required for acquiring route and configuration knowledge; however, high environmental immersion can be achieved using them.

2.5.3 Environmental Functions Planning

We call the common spatial activity of designing human operations in the environment *environmental functions planning*. For example, these functions may be organizing daily family transport, planning the functions of a farm, or designing a sports event. Environmental functions planning typically necessitates a good understanding of the spatial structure and transport networks of the environment to get multiple actors to function in a compatible way. To this end, geopictures can be very useful, as multiple simultaneous actions can be visualized in large areas, which would not be possible in the real physical environment. Route and configuration knowledge are necessary for environmental functions planning so as to plan for the required locomotion, and landmark knowledge is beneficial for conceptualizing the environment efficiently (see Table 2.1).

2.5.3.1 Terrestrial Photo Panoramas

Since multiple terrestrial photo panoramas are needed for perceiving spaces larger than a single vista, this type of geophoto would only weakly meet the requirements of environmental functions planning, as landmark knowledge has a minor role, whereas the importance of route and configuration knowledge is elevated. However, terrestrial photo panoramas may provide a planner with essential environmental details with regard to the designed activity.

2.5.3.2 Oblique Aerial Photo

Oblique aerial photos should allow useful spatial knowledge to be used in environmental functions planning due to each knowledge type being easy to acquire. Landmarks can be identified due to a high number of visible vertical features, route segments can be perceived between landmarks, and spatial configurations are feasibly conceivable in the overall picture. However, oblique aerial photos are typically produced at such high ground resolutions that their extent may not always be sufficient for representing large areas in single photos. In these cases, browsing software is necessary, which causes additional cognitive load related to steering the user interface and mentally processing the moves and rotations of the view.

2.5.3.3 Aerial Orthophoto

As configuration knowledge is the primary knowledge type for environmental functions planning, aerial orthophotos should be well suited to supporting this activity. Only partial route and landmark knowledge may cause problems for detail-dependent sub-activities of planning, such as identification of thin road barriers, but this is a minor deficit compared with having the ground surface coherently depicted.

2.5.3.4 Suitability Summary

Both aerial orthophotos and oblique aerial photos should provide firm support for carrying out environmental functions planning by covering relatively large extents in single views and allowing the easy acquisition of route and configuration knowledge. We find terrestrial photo panoramas to be less suited to this complex activity due to the arduous nature of collecting spatial information for building an understanding of the spatial structures of the environment.

2.6 Needs for Future Research

The previous sections have raised the issue of the weak direct empirical foundation in the literature about the capabilities of geophotos for building accurate and reliable human spatial knowledge and so supporting varied kinds of human spatial activities. This issue has been recently raised by other researchers, too (e.g., Bianchetti & MacEachren, 2015). Considering the deficiencies brought forth by this chapter, future research should conduct careful experimentation on how successful people are in building spatial knowledge and carrying out spatial activities when using different kinds of geophotos as the source of spatial information. There is an obvious shortage of experiments, not only related to geophoto types as such; particular scarcity is seen in studies about the use of geophotos with browsing software, although mobile applications, web maps, and other kinds of browsing software are the most usual way to view and read geophotos in the digital era.

Specific geophoto types in need of experimentation would be oblique aerial photos and surprisingly, aerial orthophotos, for which it is also hard to find empirical evidence about characteristics of use. Terrestrial photo panoramas have been involved in more studies in recent mobile navigation research, but also, their use with browsing software lacks experiments—although numerous virtual reality studies can offer plausible clues about the effects of browsing. The ongoing emergence of photorealistic and photographic virtual environments stresses the need to study the use of collated street photos in browsing software.

Experiments in which geophotos have been explicitly studied have focused mainly on wayfinding activities. Future research could thus pick topics from other kinds of spatial activities, such as the common ones presented in this chapter, which also have high importance in the lives of people and communities.

Overall, studying human use and experience of geophotos is necessary, as these photos importantly build our view and understanding of the surrounding physical world—more and more, day by day. Intensive efforts have been put into building the techniques and infrastructures that make different kinds of geophotos and their browsing software possible, so effort could also be put into understanding and developing the use of these technologies in an optimal way for people. This would certainly support the development of the technologies further still, toward more efficient, beneficial, and enjoyable use.

2.7 Summary

This chapter analyzed qualities of geospatial photographs as sources of information for constructing human spatial knowledge. The qualitative assessment provided in this chapter began with a review on the history of research on spatial cognition and spatial knowledge and an introduction to our previously developed assessment framework for geospatial pictures in general. This framework was then employed for itemizing the support of the three types of geophotos for the acquisition of the three types of spatial knowledge, and overall usefulness assessments of the geophotos were carried out. Furthermore, the support of these geophoto types was assessed for three common types of human spatial activities to describe their usability.

Finally, research needs on the topic of spatial knowledge from geophotos were identified based on the observations of the chapter.

In general, the photorealism of geophotos was assessed, both as an advantage due to their easy interpretation because of the resemblance to human visual experience and as a disadvantage due to the incoherent availability of information because of the occlusion of some features behind others. Terrestrial photo panoramas, oblique aerial photos, and aerial orthophotos were all proposed to be beneficial for providing the reader with landmark knowledge, although terrestrial photos provide the most detailed vision on landmarks. On the other hand, landmarks may be difficult to recognize in orthophotos. Route knowledge was also assessed as being easy to acquire from the three geophoto types, but most promisingly from oblique aerial photos that depict both landmarks and their interconnections on the same image. Oblique aerial photos and aerial orthophotos were assessed as successfully supporting configuration knowledge because of their aerial viewpoint; oblique aerial photos showed slight superiority due to the increased recognizability of landmarks. Terrestrial photo panoramas may make the acquisition of configuration knowledge difficult even with good browsing software.

In our qualitative assessment of geophoto types for use in common geospatial activities, oblique aerial photos also appear the most promising for wayfinding and free environmental exploration tasks. Their capability to provide the reader with the three types of human spatial knowledge can be a great advantage when using oblique aerial photos for these activities. For environmental functions planning, aerial orthophotos may provide similarly useful knowledge, or sometimes even more useful, because landmark knowledge has less importance, and an aerial orthophoto possesses nearly constant scale throughout the terrain and may contain less occlusion.

Published experimentation on the spatial knowledge from and the usability of geophotos is too sparse to fully confirm the propositions made in this chapter, and also, many other user-centered aspects of geophotos remain understudied. We believe that the importance of geophotos for human spatial knowledge acquisition and activities is very high and that these user-centered aspects deserve to be studied in the future to develop a better understanding of human use of geophotos. This better understanding can then lead to the enhanced production of geophotos with higher standards.

Acknowledgment

The author is grateful to the editors for their dedicated, unsparing, and very beneficial work in revising the chapter.

References

Beeharee, A. K., & Steed, A. (2006). A natural wayfinding—exploiting photos in pedestrian navigation systems. In *Proceedings of the 8th Conference on Human-Computer Interaction with Mobile Devices and Services* (pp. 81–88). New York: ACM.

Bianchetti, R. A., & MacEachren, A. M. (2015). Cognitive themes emerging from air photo interpretation texts published to 1960. *ISPRS International Journal of Geo-Information*, 4, 551–571.

Boren, M. T., & Ramey, J. (2000). Thinking aloud: Reconciling theory and practice. *IEEE Transactions on Professional Communication*, 43, 261–278.

Brosset, D., Claramunt, C., & Saux, E. (2008). Wayfinding in natural and urban environments: A comparative study. *Cartographica*, 43, 21–30.

Çöltekin, A., Francelet, R., Richter, K.-F., Thoresen, J., & Fabrikant, S. (2017). The effects of visual realism, spatial abilities, and competition on performance in map-based route learning in men. *Cartography and Geographic Information Science*, published online: https://doi.org/10.1080/15230406.2017.1344569.

Denis, M. (1997). The description of routes: A cognitive approach to the production of spatial discourse. *Cahiers de Psychologie Cognitive*, 16, 409–458.

Downs, R. M., & Stea, D. (1977). *Maps in Minds. Reflections on Cognitive Mapping*. New York: Harper & Row.

Fontaine, S. (2001). Spatial cognition and the processing of verticality in underground environments. In D. R. Montello (Ed.), *Spatial Information Theory. Foundations of Geographic Information Science* (pp. 387–399). Berlin, Germany: Springer.

Golledge, R. G. (1999). Human wayfinding and cognitive maps. In R. G. Golledge (Ed.), *Wayfinding Behavior. Cognitive Mapping and Other Spatial Processes* (pp. 5–45). Baltimore, MD: The Johns Hopkins University Press.

Haynes, K., Barclay, J., & Pidgeon, N. (2007). Volcanic hazard communication using maps: An evaluation of their effectiveness. *Bulletin of Volcanology*, 70, 123–138.

Hegarty, M., Smallman, H. S., & Stull, A. T. (2012). Choosing and using geospatial displays: Effects of design on performance and metacognition. *Journal of Experimental Psychology: Applied*, 18, 1–17.

Hile, H., Vedantham, R., Cuellar, G., Liu, A., Gelfand, N., Grzeszczuk, R., & Borriello, G. (2008). Landmark-based pedestrian navigation from collections of geotagged photos. *Proceedings of the 7th International Conference on Mobile and Ubiquitous Multimedia* (pp. 145–152). New York: ACM.

Hoarau, C., 2015. *Représentations cartographiques intermédiaires. Comment covisualiser une carte et une orthophotographie pour naviguer entre abstraction et réalisme?* (Unpublished doctoral dissertation.) Université Paris-Est.

Hoffman, R. R., & Markman, A. B. (Eds.) (2001). *Interpreting Remote Sensing Imagery. Human Factors*. Boca Raton, FL: CRC/Lewis.

Ishikawa, T., & Montello, D. R. (2006). Spatial knowledge acquisition from direct experience in the environment: Individual differences in the development of metric knowledge and the integration of separately learned places. *Cognitive Psychology*, 52, 93–129.

Janzen, G., & van Turennout, M. (2004). Selective neural representation of objects relevant for navigation. *Nature Neuroscience*, 7, 673–677.

Kettunen, P. (2014). Analysing landmarks in nature and elements of geospatial images to support wayfinding. (Doctoral dissertation.) Aalto University School of Engineering. Publications of the Finnish Geodetic Institute, No: 155, Kirkkonummi. Available at Aaltodoc: http://urn.fi/URN:ISBN:978-951-711-312-0.

Keuth, H. (1976). Verisimilitude or the approach to the whole truth. *Philosophy of Science*, 43, 311–336.

Liu, A. L., Hile, H., Borriello, G., Brown, P. A., Harniss, M., Kautz, H., & Johnson, K. (2009). Customizing directions in an automated wayfinding system for individuals with cognitive impairment. *Proceedings of the 11th International ACM SIGACCESS Conference on Computers and Accessibility* (pp. 27–34). New York: ACM.

Lokka, E. I., & Çöltekin, A. (2017). Toward optimizing the design of virtual environments for route learning: Empirically assessing the effects of changing levels of realism on memory. *International Journal of Digital Earth*, published online: https://doi.org/10.1080/17538947.2017.1349842.

Lynch, K. (1960). *The Image of the City*. Boston, MA: The MIT Press.

Meijer, F., Geudeke, B. L., & van den Broek, E. L. (2009). Navigating through virtual environments: Visual realism improves spatial cognition. *CyberPsychology & Behavior*, 12, 517–521.

Montello, D. R. (1993). Scale and multiple psychologies of space. In A. U. Frank & I. Campari (Eds.), *Spatial Information Theory. A Theoretical Basis for GIS* (pp. 312–321). Berlin, Germany: Springer.

Montello, D. R. (2005). Navigation. In P. Shah & A. Miyake (Eds.), *The Cambridge Handbook of Visuospatial Thinking* (pp. 257–294). Cambridge, UK: Cambridge University Press.

Münzer, S., Zimmer, H. D., & Baus, J. (2012). Navigation assistance: A trade-off between wayfinding support and configural learning support. *Journal of Experimental Psychology: Applied*, 18, 18–37.

Newcombe, N. (1985). Methods for the study of spatial cognition. In R. Cohen (Ed.), *Development of Spatial Cognition*. Hillsdale, NJ: Erlbaum.

O'Keefe, J., & Nadel, L. (1978). *The Hippocampus as a Cognitive Map*. Oxford, UK: Clarendon.

Oulasvirta, A., Estlander, S., & Nurminen, A. (2009). Embodied interaction with a 3D versus 2D mobile map. *Personal and Ubiquitous Computing*, 13, 303–320.

Passini, R. (1984). Spatial representations, a wayfinding perspective. *Journal of Environmental Psychology*, 4, 153–164.

Plester, B., Richards, J., Blades, M., & Spencer, C. (2002). Young children's ability to use aerial photographs as maps. *Journal of Environmental Psychology*, 22, 29–47.

Rehrl, K., Leitinger, S., Gartner, G., & Ortag, F. (2009). An analysis of direction and motion concepts in verbal descriptions of route choices. In K. S. Hornsby, C. Claramunt, M. Denis & G. Ligozat (Eds.), *P Spatial Information Theory. 9th International Conference, COSIT 2009* (pp. 417–488). Berlin, Germany: Springer.

Richardson, A., Montello, D., & Hegarty, M. (1999). Spatial knowledge acquisition from maps and from navigation in real and virtual environments. *Memory & Cognition*, 27, 741–750.

Richter, K.-F., & Winter, S. (2014). *Landmarks. GIScience for Intelligent Services*. Cham, Switzerland: Springer.

Sargolini, F., Fyhn, M., Hafting, T., McNaughton, B. L., Witter, M. P., Moser, M.-B., & Moser, E. I. (2006). Conjunctive representation of position, direction, and velocity in entorhinal cortex. *Science*, 312, 758–762.

Siegel, A. W., & White, S. H. (1975). The development of spatial representations of large-scale environments. *Advances in Child Development and Behavior*, 10, 9–55.

Smagorinsky, P. (1998). Thinking and speech and protocol analysis. *Mind, Culture, and Activity*, 5, 157–177.

Thorndyke, P. W., & Hayes-Roth, B. (1982). Differences in spatial knowledge acquired from maps and navigation. *Cognitive Psychology*, 14, 560–589.

Tolman, E. A. (1948). Cognitive maps in rats and men. *The Psychological Review*, 55, 189–208.

Trowbridge, C. C. (1913). On fundamental methods of orientation and "imaginary maps." *Science*, 38, 888–897.

van der Ham, I. J. M., Faber, A. M. E., Venselaar, M., van Kreveld, M. J., & Löffler, M. (2015). Ecological validity of virtual environments to assess human navigation ability. *Frontiers in Psychology*, 6, 637.

Waller, D., Hunt, E., and Knapp, D. (1998). The transfer of spatial knowledge in virtual environment training. *Presence: Teleoperators & Virtual Environments*, 7, 129–143.

3

Intersectional Perspectives on the Landscape Concept: Art, Cognition, and Military Perspectives

Raechel A. White

CONTENTS

KEYWORDS: *art, cognitive science, landscape, military*

3.1 Introduction

Human intrigue with the aerial view is lengthy. Evidence of this arises in the earliest cartographic endeavors in *Geographia*, written around AD 150. The roles that landscapes, humans, and representations play in the conceptualization of space have been taken up as a topic of study by artists, historians, psychologists, and geographers, among others. Each profession's unique viewpoints on the relationships between viewpoint, distance, and perception related to the landscape representation paint a fascinating picture of landscape perception. Landscape perception, especially the pictorial representation of landscapes through aerial photographs, has taken on immense importance, as these images are used for communication with non-expert audiences about landscape management schemes in the 19th and 20th centuries. In addition to purely communicative uses, aerial and satellite images may also serve as a means for members of the general public to contribute to scientific and management processes (Kerle & Hoffman, 2013). The development of a framework that integrates the various perspectives on landscape imagery could improve interdisciplinary discussions about how humans see the world through aerial imagery.

This current chapter provides a preliminary framework that integrates concepts across art, geography, and cognitive science about the aerial view. A number of themes permeate through these discussions that could inform further research concerning human ways of seeing the world from above.

3.2 Cognitive Geographic Information Science

Two approaches within the cognitive geographic information science (GIScience) community have emerged in relation to human image use. First, perceptual studies have addressed a number of questions concerning how humans visually scan images and use perceptual cues, such as the image interpretation elements, to identify image objects. Research has shown that photographic properties of resolution (Battersby, Hodgson, & Wang, 2012), scene context (Hodgson, 1998; Lloyd & Hodgson, 2002), and texture (Hodgson & Lloyd, 1986) also influence human interpretation. Second, a number of studies have emerged that address how knowledge and expertise affect image interpretation tasks. For example, levels of expertise, as well as different types of related expertise, have been studied in relationship to the landscape. The body of work reviewed here emerges from the sub-disciplines of GIScience dealing with cognitive aspects of working with imagery of the earth's surface from above.

Cognitive approaches to understanding the interpretation of remotely sensed images (space and aerial-borne platforms) have matured in conjunction with the technologies and techniques of photographic interpretation. In their chapter "Fundamentals of Photo Interpretation," Rabben, Chalmers, Manley, and Pickup (1960) speak to a body of work on human factors of interpretation being undertaken within the U.S. Air Force Personnel and Training Research Center to improve trainee selection. The topics of interest included logical reasoning ability, problem solving, and object identification, among others. Rabben suggests that the future of psychological research in photo interpretation should be driven by a need to better understand human performance in response to photographic media. The topics of study he suggests are visual search, identification, labeling, and problem solving, while environment, motivation, fatigue, and knowledge of results may impact human interpretation abilities.

A number of studies have been taken up since Rabben's call to arms in 1960. Perception-oriented research concerning imagery evaluates the visual characteristics of imagery in relationship to task success. For example, Battersby and Hodgson (2012) addressed the question of optimal image resolution for supporting damage assessment. An effective emergency response depends on the timely acquisition of high-resolution imagery; however, the amount of time required to acquire and process images increases with resolution. Determining the optimal balance between resolution and interpretability is useful for keeping acquisition and processing times low, thus potentially improving response times.

In other studies, the emphasis is on the comparisons between different levels of visual realism. In one such study, experiments were carried out to determine the influence of natural, or realistic, scenes on visual search processes in contrast to map search, where the background is uncluttered (Lloyd, 1997). The results of this study suggest that more complex or cluttered backgrounds and land use type influenced search time. Lloyd (1997) used feature integration theory to explain the "pop-out" effect during spatial searches. He noted that the "pop-out" effect had a greater impact on search tasks when targets were differentiated from distractor symbols based on color and combinations of color with other perceptual elements. In addition to the perceptual elements, location proved to have a strong influence on the search time.

In a series of experiments addressing the role of context in land use classification, Hodgson (1998) manipulated the window size for aerial images used to classify three Anderson Level II land use classes by undergraduate students. The results indicated that for Level II classes, window sizes of at least 40 pixels were required when analyzing imagery with spatial resolutions of around 1.5 m. Following up on that study, Lloyd, Hodgson, and Stokes (2002) conducted another set of interpretation experiments to evaluate the impact of land use class specificity and view windows on classification accuracy. The results indicated that accuracy in classification increased with more generalized land use classes and with larger view windows.

Antrop and Van Eetvelde (2000) describe how the holistic characteristic of human perception lends itself to the description of urban landscapes in remote sensing images. The authors use ecological theory of holism and Gestalt theory, from cognitive psychology, to describe human perception of landscape patches. Comparing human interpretations of landscape heterogeneity with automated algorithm approaches, the authors found that the landscape metric of "summed entropy" most closely aligned with the delineations made by human interpreters. Their results indicate the potential to develop automated methods of patch mensuration that reflect human conception of geographic space.

Dong, Liao, Roth, and Wang (2014) addressed the enhancement of maps to improve the visual search by users of maps that use satellite and aerial imagery for their base map. The imagery was enhanced through the use of filters and stretches common to remote sensing applications. Participants were instructed to identify as many roads as possible from the imagery, either an unenhanced version or an enhanced version of the image, in 8 s. The application of image enhancement techniques improved both the effectiveness and the efficiency of target identification, particularly in areas with complex landscapes.

In a series of studies examining the role of shadow in interpretation of relief features, Çöltekin and others examined how personal traits and direction affect the perceptual illusion of relief inversion. Relief inversion, the process by which shadow orientation deceives an interpreter's perception, was a commonly discussed topic among early authors on interpretation, and studies were carried out by interpreters, a number of which are identified in Rabben et al. (1960).

The first study in the series, by Poveda and Çöltekin (2014), evaluated the prevalence of relief inversion in both imagery sets as well as in the general population. In the user study portion of this paper, 535 subjects completed a series of interpretations and relative height judgments from imagery representing a variety of terrains and land covers. The results of this survey, although preliminary, indicated that 180° rotation of images does reverse the perception of imagery for the majority of participants, and it is most relevant in complex topographic landscapes. In their exploratory analysis of group differences, they found some evidence that expertise, or experience with imagery, improves accuracy.

In a subsequent study by Biland and Çöltekin (2017), an empirical assessment of the effects of shadow on shape perception was performed. In this study with shaded relief maps, the authors again found that an increase in experience with imagery improved accuracy in landform identification. The results of this second study also indicated that males, especially untrained males, exhibited overconfidence in their interpretations. Finally, the study found that landform interpretation accuracy was highest at 337°, supporting the general belief that the best illumination point is from the northwest corner of an image, but contrasting with the optimal 31° illumination from cartographic convention. Biland and Çöltekin (2017) also found that experts experience the pseudopsychological shadow illusion markedly less with the images (whereas this difference is not obvious for the shaded relief maps), which, they argue, might be because the cognitive processing overrides the perceptual signal. These findings build on previous work by Davies, Tompkinson, Donnelly, Gordon, and Cave (2006) and Henderson, Brockmole, Castelhano, and Mack (2007), which showed that visual saliency was a more important factor than scene context in the search processes of experts and novices due to image complexity. The studies described here and earlier reflect the long tradition of cognitive research, both empirical and informal, regarding the perceptual influences on image interpretation. They also highlight the fact that many topics that were of interest in the early 20th century continue to be of interest to today's scientists.

The studies described earlier contrast markedly with studies in which the influence of expertise is the main concern. Medin, Lynch, Coley, and Atran (1997) compared the knowledge structures of expert taxonomists, landscape workers, and parks maintenance personnel using a card sorting method and found that the underlying goals of an individual's domain, and thus, their own motives, greatly affected their category structure.

Evaluating the relationship between perception and experience, Lansdale et al. (2010) used eye-tracking experiments to evaluate expert-novice differences in aerial image change detection tasks. They found that an object's visual saliency was a more important factor during novice search than during expert search, suggesting that the prior knowledge of expert analysts prevails over salience in interpretation tasks. Previous work performed by Lloyd et al. (2002) examined the effect of expertise on land use categorization with aerial imagery. Expert geographers tended to respond more quickly and have higher confidence than non-experts overall, and experts were more accurate in the identification of low-level land use categories such as "industrial areas." Battersby et al. (2012) found that experts in either hazards management or image interpretation were more accurate than novices at damage assessment tasks. These authors suggest that the amount of improvement on accuracy as a function of expertise is limited, but that expertise may have a stronger effect in more complex image interpretation tasks.

Hoffman (1985) analyzed a database of landscape feature propositions and predications used by expert terrain analysts. Analysis of the database led to the development of a corpus of terms for identifying and describing landforms. The majority of generic propositions and predications were found to be further qualified with adjective descriptors. This work was then used to inform expert system design. In later work, he also developed a list of terrain analysis terminology drawn directly from the perceptual descriptions of terrain features by expert analysts (Hoffman, 1990). This work reflects a trend in cognitive remote sensing research where expert systems or knowledge-based systems design was the central focus. Today, these areas of work serve as the roots of recent research in geographic object-based image analysis.

Gardin et al.'s (2011) set of digitization studies addressed the quantification of interpreter variability. Analysts were asked to perform six separate digitizations of features in imagery, including tasks using lines, polygons, and points. The completeness of human interpretations ranged from 11% to 100% between individuals, and accuracy ranged from 35% to 100% between individuals, depending on object geometry. Human factors contributing to this variability included aspects of user confidence, desire to achieve high accuracy, and previous experience (Van Collie et al., 2014).

Hand in hand with expertise is the concept of visual literacy, one's ability to interpret and extract meaning from images. A number of scholars have evaluated visual literacy in school children related to aerial imagery (for early work see Dale [1971]). Svatoňová (2017) examined students' ability to perform visual tasks and the potential to use these tasks for improving student visual literacy. She used Image-Reader in her study and evaluated the interpretability of landscape changes in natural color images, false color images, and maps by school students between the ages of 11 and 19 years. Age was found to be a significant factor influencing their ability to detect change in images. This task was found to be appropriate only for students in the 15–19-year-old categories. The results also showed that false color imagery was interpretable by students of all ages.

The goal of many of the cognitive GIScience studies is the improvement of information extraction from earth imagery. To improve insight generation from imagery, researchers study the effects of image qualities, task goals, and human factors. Studies have indicated that expertise can improve the accuracy and confidence of interpretation and that types of expertise impact the interpretations that result from the use of images. Additionally, image quality factors, such as resolution, scale, and context, can greatly impact the interpretation results. Much of the work generated by geographic information scientists has built on earlier studies arising from military research concerning aerial reconnaissance during the first half of the 20th century.

3.3 Reconnaissance

As with many technological advancements, some senior officials showed skepticism about aerial photography's effectiveness for military engagement (Campbell, 2008; Winchester & Wills, 1928). Skepticism was overcome by the successes at the front lines by the British and French air forces (Smith, 1958). The practices of image capture and interpretation were new at this point and evolving at a fast pace. Several books and texts emerged at the end of the war, such as the U.S. War Department's pamphlet *Study and Exploitation of Aerial Photographs* (1918). With the onset of World War II, literature began to proliferate from the U.S. War Department concerning air photo interpretation, the focus of these pamphlets being methods of interpretation and sanctioned means for information gathering for military reconnaissance. The main goal of these texts was to build an interpreter's knowledge of the defensive strategies of the enemy, including topics on troop movement, camouflage use, and the spatial layout of the battlefield.

To be successful at interpretation, it was believed that an interpreter would need meticulous education, experience, and natural aptitude (Bianchetti & MacEachren, 2015). Viewing aerial photographs became a constructed task whereby the image content was visually dissected using stringent guidelines based on the early experiences of World War I and World War II interpreters. Records of image interpretation results were kept to inform decision making in cases where the task was difficult, and scrapbooks were advocated as a means for recording unique cases (Lobeck & Tellington, 1944). To achieve consistent results among various interpreters, the military developed a systematic means of analyzing these images.

Descriptions of the interpretation process within texts written during the early 20th century focused on the use of perceptual cues. In "Elements of Photographic Interpretation Common to Several Sensors" (Olson, 1960), the perceptual cues of shape, shadow, size, tone, texture, pattern, and resolution were used to construct a systematic method of building evidence for or against the presence of various objects, such as tanks, fences, railroads, and enemy troops. For example, shadows could be used to determine the shape, size, and in some cases, even the presence of an object. "Shadows may disclose the shape and size of an object even though the object itself may be unrecognizable on a vertical photograph [p. 75] (Eardley, 1942)." To create knowledge beyond simple identification of objects, it was necessary for analysts to infer meaning through the combination of image interpretation elements via the process of convergence of evidence, conscious deductive analysis of images, and ancillary data to determine the identification of image objects and the signification of their importance (Colwell, 1993). The dominant image interpretation elements used for these analytical tasks, according to texts from that time, were tone and shadow. These elements provided clues to analysts about the shape and size of objects on the ground, while the analyst's domain knowledge and personal experience provided them with the ability to infer greater meaning about the landscape.

Later, as a standard method of interpretation was adopted, research turned toward training improvements. A great body of work (completed within the military) was developed, though most of it is difficult to access. For example, Townsend and Fry (1960) evaluated the potential for improving novice free-search patterns through training. With the aid of a digital guide, their free-search patterns could be improved. In an eye-tracking study of three untrained participants, Townsend and Fry (1960) evaluated the effect of scan speed and object contrast on the ability of subjects to visually track a marker over aerial photographs. The authors found that in cases where contrast was acceptable (10% or higher), free scanning by interpreters was faster than automatic guided scanning methods. However, in cases with lower contrast (6% and lower), failure rates increased to 100% during free search, whereas guided search increased the successful identification of the target (although it was slower than free search). Evaluating the rate of eye movements per scan line, the authors found that contrast had no effect, despite the fact that previous free search studies had shown eye fixation to increase as contrast decreased.

Aerial reconnaissance through direct observation was also important to military strategizing. Studying the relationship between aircraft speed, distance, and object contrast, Duntley (1953) used model-size objects of interest (toy soldiers and military equipment) and moving automobiles to evaluate the relationships between visual contrast, aircraft speed, and height. From these modeled runs, Duntley determined the maximum altitude at which object detection and recognition could occur. Additionally, Duntley found that object recognition success varied with visual search and recognition methods. Those achieving the highest accuracy "do not watch the target continuously" but instead, "their eyes search other parts of the scene but return to the target for brief, steady views." He went on to suggest that the time period between detection and recognition approximates to 2.7 s, although there was no direct measurement of this time interval.

The Human Factors Research branch of the Adjutant General's Research and Development Command, U.S. Army, conducted a number of research studies concerning image interpretation during the 1960s. In an address of the Photogrammetric Engineering Society in 1961, Dr. Robert Martinek summarized their work, identifying a number of analytical tasks associated with the military applications. At the time, new types of images were beginning to emerge, altering the contemporary understanding of image interpretation within the U.S. Army. Martinek and Sadacca (1961) emphasized the importance of improving human work to overcome the limitations of imagery quality.

The first of the pilot studies described by Martinek and Sadacca (1961) compared the accuracy and correct identification of truck convoys from different visual perspectives (nadir, oblique, and nadir + oblique). The results of this study indicate that providing interpreters with both nadir and oblique viewpoints did not improve their interpretation accuracy or identification correctness. In the next experiment in this series, stereo and non-stereo viewing conditions were compared. Again, participants were provided with different images of varying modes of display (stereo or non-stereo) and asked to identify tactical objects from the images. Similarly to the study evaluating the use of visual perspective, this study found no statistically significant differences in correctness between the two image types. The final study reported here gauges the impact of intelligence reports on expectancy during interpretations. Groups of interpreters were either given no intelligence or given intelligence that would indicate the importance of some target, such as hospitals. The results indicated that a far larger number of targets were identified in the situation where intelligence information was provided, but that it impacted those interpretations only in cases where other factors, such as object ambiguity, might impact the results.

The literature reviewed in this section is representative of a much larger body of work that emerged from the U.S. military regarding human factors of image interpretation. The stated goal for these works is most often the improvement of tactical and strategic techniques employed by photographic interpreters. Through manipulation of the working environment, these studies could make recommendations for the refinement and improvement of interpretation methods, including the inclusion of additional knowledge sources, and impacts on interpretation due to image characteristics. Systematism in the interpretation of remotely sensed images led to the development of an alternative way of viewing reality, beyond the aerial viewpoint. As Litfin (1997) suggests, the systematic image interpretation processes developed by military officers in World War I, and to a greater extent in World War II, forced the interpreter to break down the image into visual clues that could inform their conjectures about ground conditions. The systematic method of deconstructing imagery into intelligence that was developed during this time period complements the unique ways of viewing landscapes from the perspective of artistic critique.

3.4 Art

Discourse around the landscape in art has an equally important perspective to provide on viewing and understanding the earth's landscapes. From a critical standpoint, aspects of image production, aesthetic, and sociocultural context can be examined. Previous scholars have often recognized a connection between the process of aerial photography and interpretation with artistic endeavors of production and viewing. Here, connections between the conception of landscape, representation, and the aerial view are made to provide alternative perspectives on the relationship between viewer, landscape, and sensor.

Landscape painting has wrestled with the relationship of viewpoint, perception, and distance for quite some time. Originating from the Dutch "Landschap," the traditional landscape painting is characterized by a large tract of land, often in a panoramic view. It has been suggested that the core of landscape painting is the composition of space (Cosgrove, 1985). In the 16th century, these paintings used the landscape as a frame around human activity. By the 17th century, classical landscapes had become contrived and formulaic representations, the landscape serving as an idealistic stage for human presence. The points of view used in such paintings suggest a subjection of the landscape by the viewer.

Early 20th-century art movements such as Cubism and Futurism have also found a connection with aerial photography. Qualities common across Cubism and aerial photographs include the reduction of depth, reduction of detail, and composition of simple geometric shapes. These qualities contrast with the complex, detailed, even chaotic perspective that one experiences when in direct proximity to the landscape. Saint-Amour (2003) argues that parallel developments in military aerial reconnaissance and the avant-garde led to the development of similar new methods of interpretation characterized by systematic, controlled viewing.

As Amad (2012) points out, the dichotomy of the "god view" from above and the "man view" from below has pervaded cartographic and artistic representations for centuries. Linking the "god view" to experience, one finds that there is a dehumanizing quality in the abstraction and panoptic view it affords. Starting with landscape painting, a genre of art that was overlooked for a lengthy amount of time, we review the role that these representations have had in evoking a response from viewers.

The earliest photograph, *View from the window at Le Gras* by Niepce in 1826, was taken from an elevated position from a window overlooking several buildings. The French photographer Gaspard Felix Tournachon, commonly known as Nadar, took to the skies to capture the first photographs of Paris from hot air balloons in 1958. While his first image has long been lost, his textual accounts of his aerial photography adventures have remained. These texts provide evidence of his curiosity regarding the knowledge to be gained from the aerial photographs that he captured (Bann, 2013). He foretold the possibility of using aerial photography for both mapping and military purposes (Mifflin, 2016). As Bann (2009) points out, as is evident from Nadar's "Photography the homicide," Nadar was well aware of the potential social implications that photography could have on society's moral compass.

Of course, not all artistic representations of the landscape are by skilled painters. The notable balloon flights of Thomas Baldwin in Chester, U.K. are an example of unifying art and science in viewing the earth (Thebaud-Sorger, 2014). In his own retelling of his first impressions of the landscape from above, Baldwin speaks of the sublimity, grandeur, and beauty of the landscape in his first views of the Chester landscape. He describes in detail the scenes that he witnesses, including their scale, shadow, color, and arrangement. He compares the scale of the scene to those of "Burdett's Map" and the city of Lilliput from *Gulliver's travels*. The flattened perspective that he describes is echoed in aerial photography texts in the 20th century. His amazement over the scene is reflected in his colorful descriptions of it. His sketches of his experience are equally telling. His exposition on this experience serves as a reminder of how impactful the aerial view was on the perceptions of the earliest aeronauts (Thebaud-Sorger, 2014).

Several common themes emerge from these artistic practices. The first is the idea of point of view or perspective. In landscape photography, Cubism, and even the earliest sketches from hot air balloons, the vertical perspective is seen as a means for expanding the scope of human vision. While not necessarily vertical, the viewpoint of many of the landscape paintings is also wide in scope and gives an elevated view of the landscape scene. Second, in classical landscape photography and landscape painting, there is an emphasis on the importance of realism. This realism is explained in part by the level of detail the media produce but also by the viewpoints they afford the viewer (Cosgrove, 1985). Third, the flattening of the earth's surface as distance from it increases is discussed in the relationship between Cubism and avant-garde art as well as the sketches and transcripts by Baldwin (1785). The relationships between these various viewpoints from art, cognitive science, and the armed forces are synthesized in the following section.

3.5 Discussion

The framework presented here is an exploration of common themes permeating various discourses on ways of seeing the earth. While not a comprehensive review of literature across the disciplines surveyed here, it is an important first step toward developing a comprehensive framework for understanding ways of seeing the world by remotely sensed imagery. The development of such a framework does two things. First, it provides a shared

TABLE 3.1

Several Topics Themes Which Emerged from the Disciplinary Readings of Art, Military Science, and Cognitive Science Related to the Landscape Perspective

	Discipline		
Characteristic	Art	Military	Cognitive Science
Viewer	Relationship to landscape	Intelligence constructor	Participant
Task	Aesthetic	Strategic/tactical	Varies
Representation	Painting, photography	Imagery	Maps, imagery
Viewpoint	Power relationship	Information	Context
Objects	Symbolism	Importance	Discernibility

conceptual model for developing integrated methods of image analysis. Second, a review intersecting art and science in this way may prompt in those tasked with viewing images a new perspective to reflect on their own work.

In Table 3.1, the characteristics of the discussions reviewed here are presented as they relate to each of the disciplines included. This framework is very much a work in progress. Medical, psychological, astronomical, and industrial ways of viewing, among others, should be evaluated to advance a model image analysis.

A general model of interpreting visual representations may be borrowed from MacEachren and Ganter (1990) or adapted from Svatoňová (2017). The three main components of any model of geographic visualization are the viewer, the image, and reality. In the case of remote sensing, an additional actor, the sensor, could also be taken into consideration. Here, the image has been generalized to represent all models of high–spatial resolution sensors.

The *viewer* in the three cases presented here differs. For the artist, the viewer is at some distance from the landscape, viewing it as an observer. For the military, the viewer was also distant, but this distance was overshadowed by the viewer's role as an intelligent agent constructing information from the landscape. Finally, in cognitive science, participants are evaluated for their performance of a given task in much the same way as military trainees were evaluated in their own work; however, their experience and their goals of interpretation may vary. It is possible that a viewer from this perspective could fulfill either of the viewer roles in the art or military descriptions.

The use of *objects* in the landscape varies dramatically between art and science. In the case of art, landscape artists tended to develop contrived scenes in which objects and relationships between objects symbolized a variety of themes, such as religion. Similarly, the objects that form aerial images also have meaning to the interpreters that use them; their symbolism, however, is more closely tied to the specific strategic or tactical goals that an interpreter is given. The objects themselves may or may not be purposely placed in the landscape. Finally, objects are typically treated as either a region of interest or a target to be identified by a participant in cognitive GIScience experiments. There may be multiple types of these targets, and they may or may not be fully discernable to the participant. The objects may or may not hold importance to them beyond the confines of the experiment.

The types of *tasks* that a human may pursue within the context of art, military, or cognitive science may vary widely. Viewing a piece of art as a consumer is often presented as an aesthetic experience. As an exception to this passive viewing, Cubism and avant-garde pieces of art often demanded a critical deconstruction of the pieces to understand the sociocultural symbolism presented by the artist. Viewing images within the context of military applications was largely for information derivation, often for tactical purposes, although the navigational use of imagery also applies. Finally, viewing imagery as a participant in cognitive science studies can range from object detection to in-depth description and interpretation for deriving information about the ways in which people see with imagery.

Just as the goal of interpretation across the three disciplines varies, so do the ways in which the landscape is represented. *Representations* of the landscape by artists vary greatly by genre. As indicated by Saint-Amour (2003), the methods of representation varied greatly between landscape artists, Cubists, and Futurists. The variation of representations used by military interpreters was less varied until the adoption of color photography in the late 1950s. Prior to that point, black and white photography dominated, and the quality of imagery was constrained by technological capabilities. Finally, cognitive GIScientists have used a wide variety of representations across a broad spectrum of abstraction levels, from two-dimensional (2D) maps to three-dimensional (3D) virtual reality systems. If we focus just on the 2D image studies presented here, image quality, including resolution, color, and scale, varies extensively across the studies.

An important aspect of the work reviewed here is the role of *viewpoint*, or perspective, in how humans see the earth. While not explicitly addressed in models of visualization, it plays a critical role in the discourse about the way we see the earth. From the artistic perspective and even the cartographic one, the nadir viewpoint has been long used as a method of establishing dominance and symbolizing the subjugation of populations. From a military perspective, the nadir and oblique viewpoints are examined in relation to the amount of information that can be extracted about a location. In early texts, the oblique viewpoint is favored over the nadir one for many applications, because it affords the viewer a more natural view of the earth's surface than nadir does. Finally, in the case of cognitive GIScience, viewpoint has been examined to a lesser extent, but more recent contributions in regard to wayfinding and spatial thinking have used alternative, typically groundbased, photographic viewpoints.

Critical reflections within these domains may also provide new insights for critiquing our own practices as geographic information scientists. There is also a need to expand this work to consider new technologies for representing the landscape, such as virtual reality. Their ability to transform the human experience of landscape, and their ability to overcome some of the challenges presented in critiques of traditional remote sensing methods, evoke new critiques on ways of seeing with geospatial technologies. For example, unmanned aerial systems, while still developing a digital, and thus pixelated, image, do so at remarkably high spatial resolutions that can *also* capture humans.

3.6 Conclusions

Viewing the landscape from above is no longer a privilege only for the military interpretation officer. Today, such imagery is used extensively to support narratives about world events in popular news media. The introduction of drone warfare has brought the aerial gaze to television sets around the world (Adey, Whitehead, & Williams, 2011). These images contrast with the typical aerial image described in the early literature. Often, such images present a highly localized picture of the environment and can leave doubts in the minds of viewers regarding their fidelity. For example, in her critique of the Operation Orchard mission images, Kaplan (2015) suggests that even experts held some reservations regarding whether the buildings shown were truly nuclear facilities. This begs the question: if experts are unsure of image contents, how can typical television viewers with little to no experience with these types of aerial images be expected to have faith in them? This framework for understanding various perspectives regarding the ways we see the earth is a first step to improving our ability to improve visual literacy among non-experts by finding common terms and concepts to start discussions regarding image understanding.

References

Adey, P., Whitehead, M., & Williams, A. J. (2011). Introduction: Air-target distance, reach and the politics of verticality. *Theory, Culture and Society*, 28(7–8), 173–187.

Amad, P. (2012). From god's-eye to camera-eye: Aerial photography's post-humanist and neo-humanist visions of the world. *History of Photography*, 36(1), 66–86.

Antrop, M., and Van Eetvelde, V. (2000). Holistic aspects of suburban landscapes: Visual image interpretation and landscape metrics. *Landscape and Urban Planning*, 50(1–3), 43–58.

Baldwin, T. (1785). *Airopaidia: Containing the Narrative of a Balloon Excursion from Chester, the Eighth of September*. London: Fletcher.

Bann, S. (2009). "When I was a photographer": Nadar and history. *History and Theory*, 48(4), 95–111.

Bann, S. (2013). Nadar's aerial view. In M. Dorrian & F. Pousin (Eds.), *Seeing from Above: The Aerial View in Visual Culture* (pp. 83–94). London: IB Tauris.

Battersby, S., Hodgson, M. E. W., & Wang, J. (2012). Spatial resolution imagery requirements for identifying structure damage in a hurricane disaster: A cognitive approach. *Photogrammetric Engineering and Remote Sensing (PE&RS)*, 78(6), 625–635.

Bianchetti, R. A., & MacEachren, A. M. (2015). Cognitive themes emerging from air photo interpretation texts published to 1960. *ISPRS International Journal of Geo-Information*, 4(2), 551–571.

Biland, J., & Çöltekin, A. (2017). An empirical assessment of the impact of the light direction on the relief inversion effect in shaded relief maps: NNW is better than NW. *Cartography and Geographic Information Science*, 44(4), 358–372.

Campbell, J. B. (2008). Origins of aerial photographic interpretation, US army, 1916 to 1918. *Photogrammetric Engineering and Remote Sensing*, 74(1), 77.

Colwell, R. N. (1993). Four decades of progress in photographic interpretation since the Founding of Commission VII (IP). *International Archives of Photogrammetry and Remote Sensing*, 29, 683–690.

Cosgrove, D. (1985). Prospect, perspective and the evolution of the landscape idea. *Transactions of the Institute of British Geographers*, 10, 45–62.

Dale, P. (1971). Children's reactions to maps and aerial photographs. *Area*, 170–177.

Davies, C., Tompkinson, W., Donnelly, N., Gordon, L., & Cave, K. (2006). Visual saliency as an aid to updating digital maps. *Computers in Human Behavior*, 22(4), 672–684.

Dong, W., Liao, H., Roth, R. E., & Wang, S. (2014). Eye tracking to explore the potential of enhanced imagery basemaps in web mapping. *The Cartographic Journal*, 51(4), 313–329.

Duntley, S. Q. (1953). The limiting capabilities of unaided human vision in aerial reconnaissance. Ann Arbor, MI: Armed Forces, National Research Council, Vision Committee Secretariat.

Eardley, A. J. (1942). *Aerial Photographs: Their Use and Interpretation*. New York: Harper and Brothers.

Gardin, S., van Laere, S. M. J., van Coillie, F. M. B., Anseel, F., Duyck, W., de Wulf, R. R., & Verbeke, L. P. C. (2011). Remote sensing meets psychology: A concept for operator performance assessment. *Remote Sensing Letters*, 2(3), 251–257.

Henderson, J. M., Brockmole, J. R., Castelhano, M. S., & Mack, M. (2007). Visual saliency does not account for eye movements during visual search in real-world scenes. In R. P. G. van Gompel, M. H. Fischer, W. S. Murray & R. L. Hill (Eds.), *Eye Movements: A Window on Mind and Brain* (pp. 537–562). Boston, MA: Elsevier.

Hodgson, M. E. (1998). What size window for image classification? A cognitive perspective. *Photogrammetric Engineering and Remote Sensing*, 64(8), 797–807.

Hodgson, M. E., & Lloyd, R. E. (1986). Cognitive and statistical approaches to texture. Paper presented at the ASPRS–ACSM, Falls Church, VA.

Hoffman, R. R. (1985). *What Is a Hill? An Analysis of the Meanings of Generic Topographic Terms*. Research Triangle Park, NC: Battelle Memorial Institute.

Hoffman, R. R. (1990). What is a hill? Computing the meanings of topographic and physiographic terms. In U. Schmitz, R. Schutz & A. Kunz (Eds.), *Linguistic Approaches to Artificial Intelligence* (pp. 97–128). Frankfurt, Germany: Verlag Peter Lang.

Kaplan, C. (2015). Air power's visual legacy: Operation Orchard and aerial reconnaissance imagery as ruses de guerre. *Critical Military Studies*, 1(1), 61–78.

Kerle, N., & Hoffman, R. R. (2013). Collaborative damage mapping for emergency response: The role of cognitive systems engineering. *Natural Hazards and Earth System Sciences*, 13(1), 97–113.

Lansdale, M., Underwood, G., & Davies, C. (2010). Something overlooked? How experts in change detection use visual saliency. *Applied Cognitive Psychology*, 24(2), 213–225.

Litfin, K. T. (1997). The gendered eye in the sky: A feminist perspective on earth observation satellites. *Frontiers: A Journal of Women Studies*, 18(2), 26–47.

Lloyd, R. (1997). Visual search processes used in map reading. *Cartographica: The International Journal for Geographic Information and Geovisualization*, 34(1), 11–32.

Lloyd, R., & Hodgson, M. E. (2002). Visual search for land use objects in aerial photographs. *Cartography and Geographic Information Science*, 29(1), 3–15.

Lloyd, R., Hodgson, M. E., & Stokes, A. (2002). Visual categorization with aerial photographs. *Annals of the Association of American Geographers*, 92(2), 241–266.

Lobeck, A. K., & Tellington, W. J. (1944). Air photographs. In *Military Maps and Air Photographs: Their Use and Interpretation* (pp. 199–234). New York: McGraw-Hill Book Company, Inc.

MacEachren, A. M., & Ganter, J. H. (1990). A pattern identification approach to cartographic visualization. *Cartographica: The International Journal for Geographic Information and Geovisualization*, 27(2), 64–81.

Martinek, H., & Sadacca, R. (1961). Human factors studies in image interpretation. *Photogrammetric Engineering*, 27(5), 714–728.

Medin, D. L., Lynch, E. B., Coley, J. D., & Atran, S. (1997). Categorization and reasoning among tree experts: Do all roads lead to Rome? *Cognitive Psychology*, 32, 49–96.

Mifflin, J. (2016). Book review: *When I Was a Photographer. Early Popular Visual Culture*, 14(3), 289–291

Olson, C. E. (1960). Elements of photographic interpretation common to several sensors. *Photogrammetric Engineering and Remote Sensing*, 26(4), 651–656.

Poveda, B., & Çöltekin, A. (2014). Prevalence of the terrain reversal effect in satellite imagery. *International Journal of Digital Earth*, 8(8), 640–655.

Rabben, E. L., Chalmers Jr., E. L., Manley, E., & Pickup, J. (1960). Fundamentals of photo interpretation. In R. N. Colwell (Ed.), *Manual of Photographic Interpretation* (pp. 99–168). Washington, DC: American Society of Photogrammetry and Remote Sensing.

Saint-Amour, P. K. (2003). Modernist reconnaissance. *Modernism/Modernity*, 10(2), 349–380.

Smith, C. B. (1958). *Evidence in Camera: The Story of Photographic Intelligence in World War II*. London: Chatto and Windus.

Svatoňová, H. (2017). Reading satellite images, aerial photos and maps: Development of cartographic and visual literacy. In P. Karvánková, D. Popjaková, M. Vančura & J. Mládek (Eds.), *Current Topics in Czech and Central European Geography Education* (pp. 187–208). Berlin, Germany: Springer.

Thebaud-Sorger, M. (2014). Thomas Baldwin's Airopaidia, or the aerial view in colour. In M. Dorrian & F. Pousin (Eds.), *Seeing from Above: The Aerial View in Visual Culture* (pp. 46–65). London: I. B. Taurius and Co. Ltd.

Townsend, C. A., & Fry, G. A. (1960). Automatic scanning of aerial photographs. Paper presented at the Symposium on Visual Search Techniques, Washington, DC.

United States War Department Division of Military Aeronautics (1918). *Study and Exploitation of Aerial Photographs.* Washington, DC: U.S. Government Printing Office.

Van Collie, F., Gardin, S., Anseel, F., Duvuk, W., Verbeke, L., & De Wulf, R. (2014). Variability of operator performance in remote sensing image interpretation: The importance of human and external factors. *International Journal of Remote Sensing*, 35(2), 754–778.

Winchester, C., & Wills, F. (1928). *Aerial Photography: A Comprehensive Survey of Its Practice and Development.* London: Chapman & Hall.

4

Head in the Clouds, Feet on the Ground: Applying Our Terrestrial Minds to Satellite Perspectives

Ryan V. Ringer and Lester C. Loschky

CONTENTS

KEYWORDS: *aerial scenes, satellite imagery, scene categorization, scene gist, scene perception*

4.1 Introduction

As satellite imagery becomes more available and detailed, we find more uses of it to improve our everyday lives. Aerial views of our world can be used to assess topography, soil composition, and the amount of photosynthetic productivity of an ecosystem and even to measure the degree to which an area is inhabited by humans. Some questions we ask of remote sensing can be answered relatively directly using simple physical attributes of the image. For instance, spectral analysis of satellite imagery can provide quick, automated estimates of photosynthetic productivity for a given plot of land or the turbidity of a stream by using normalized difference indices. However, other questions require more abstraction to give meaning to aerial scenes. For example, how do we come to the conclusion that we are looking at a beach, a desert, a city, or a forest? Humans can solve this type of problem from the ground-based perspective in a fraction of a second. Often, we refer to this type of categorization as understanding the "gist" of a scene (Biederman, Rabinowitz, Glass, & Stacy, 1974; Fei-Fei, Iyer, Koch, & Perona, 2007; Loschky & Larson, 2010; Oliva, 2005; Potter & Levy, 1969; Schyns & Oliva, 1994; Thorpe, Fize, & Marlot, 1996). Specifically, we define scene gist as the holistic semantic representation of a scene that is acquired within a single eye fixation (Larson, Freeman, Ringer, & Loschky, 2014; Loschky, Ringer, Ellis, & Hansen, 2015). In fact, this very early level of visual processing is a fundamental step in interpreting the world around us, providing the foundation on which later processes, such as visual search, can be performed with greater efficacy.

It has taken eons of evolutionary selection to come up with a visual system that is optimized in such a way that it can quickly extract and exploit our environment so seamlessly, and artificial intelligence has only recently begun to match human categorization performance. Given that our entire evolutionary history, until the last 100 years, has been spent on the ground, it is reasonable to speculate that the visual cues from terrestrial views of the earth may not translate to recognizing satellite imagery. In fact, recent research on this question confirms the fact that humans are worse at recognizing scenes from aerial views than they are for terrestrial views; however, clear similarities between aerial and terrestrial scene gist do seem to exist (Loschky et al., 2015). Thus, in terms of applications to the field of remote sensing, there is a two-fold problem, in that (1) automated semantic categorization of scenes is not yet equal to or better than human performance and (2) most computational models for categorizing scene images are not optimized for aerial views. Therefore, the best (and current) solution resides in the realm of human expertise. However, there is only a small, although growing, body of research on this topic.

From a cognitive theoretical perspective, fascinating questions are raised by the difficulties faced by human viewers of aerial views of scenes, and attempting to answer them can tell us important things about how the human mind operates and possibly also its evolutionary history. In this chapter, we will provide a brief background on typical (terrestrial) scene perception as well as a review of the available aerial scene perception literature. Furthermore, we will discuss cognitive frameworks for both expertise and evolutionary constraints on learning and their implications for aerial scene gist recognition. Finally, we will explore the practical utility of aerial scene gist as it relates to cognitive maps and navigation.

4.1.1 A Brief Introduction to Scene Gist Recognition

Our ability to categorize and understand the world around us is extremely rapid. Behavioral research has shown that people can accurately identify the category of a scene after as little as 12 ms of uninterrupted input from the retina, and asymptotic performance is observed after only 50 ms (Loschky, Hansen, Sethi, & Pydimari, 2010). Furthermore, research using electroencephalography (EEG) and magnetoencephalography (MEG) has demonstrated that brain activity elicited by scene images in a scene categorization task begins to show unique differences between scene categories from approximately 150 ms after an image has been presented, with categories being largely distinct after 200 ms (Ramkumar, Pannasch, Hansen, Larson, & Loschky, 2016; Thorpe et al., 1996). Within these fractions of a second, our brain is hard at work filtering and reconstructing the scene at different spatial scales (Hegde, 2008; Schyns & Oliva, 1994). At early stages of processing, we can get an idea of some important physical attributes of the scene, such as whether it is a natural (e.g., "forest") or man-made (e.g., "city") environment (Loschky & Larson, 2008, 2010), as well as other information such as its depth and navigability (Greene & Oliva, 2009a). This initial representation has been said to be *holistic* (Oliva & Torralba, 2006) because it does not assume the scene to be the sum of its parts or objects; rather, it is an all-encompassing structural environment—the place where objects can be found.

The relationship between the gist of the scene and the objects within it is important, because it allows us to use information gathered during the early, coarse sweep of our visual field so that we can make more informed decisions about where to look for an object. For instance, for people to avoid most large predators in the wild, they would likely pay more attention to the areas close to the ground instead of up in the air. In modern-day life, we still rely on these probabilistic cues to guide our attention. For instance, as a pedestrian crossing an intersection, you will be much more successful at avoiding oncoming traffic if your attention is allocated to the street rather than the sidewalk. Thus, the relationship between scene structure and semantics has been a defining force for our visual system. Our ability to scan and navigate the earth from an aerial perspective has nowhere near the level of speed and accuracy that it has from our ground-based perspective, but at the same time, aerial images are far from unrecognizable, as seen in Figure 4.1. The fact is, we simply do not process them with the same level of efficiency. So, what is it that makes these viewpoints so different, and how does that affect the way we can interpret satellite views?

4.2 Scene Perception from a New Perspective: A Case for Scene Perception as Visual Expertise

The first question we must address is *what* critical information is missing from aerial scenes. One critical structural component of terrestrial scenes is the presence of a horizon—the division between the sky and ground, which

FIGURE 4.1
Example images from a scene categorization task. Each example scene image was chosen based on having been recognized in a study at the mean accuracy level for its respective category and view (aerial or terrestrial). Images were modified so that all had the same mean luminance and contrast of the entire image set. (From Loschky, L.C., et al., *J. Vis.*, 15(6:11), 1–29, 2015.)

produces a dominance of horizontally oriented visual structure in real-world natural scenes (Hansen & Essock, 2004). The unique, semantic distinction between sky and ground contributes to an important perceptual anchor, called the *perceptual upright* (Haji-Khamneh & Harris, 2010; Harris, Jenkin, Dyde, & Jenkin, 2011), namely, "the orientation at which objects appear the right way up" (Harris et al., 2011, p. 135), or the *gravitational frame*, which is a constraint on visual perception based on "the fact that we usually view scenes with our bodies aligned vertically with gravity and our head above our feet" (Loschky et al., 2015, p. 2). We unconsciously assume that when we step outside, the sky will be above us and the earth will be at our feet. But what would happen to our ability to recognize the gist of a scene if we lost this critical feature of our environment? How useful are our assumptions about the world when they are divorced from such an implicit perceptual landmark? This is the situation faced when recognizing an aerial view of a scene, in which there is no horizon, a situation missing from our evolutionary history until only very recently. This could likely cause difficulties in scene recognition, at which humans can be argued to be experts, because theories of expert image recognition argue that our ability to rapidly recognize targets of expertise with high degrees of speed and accuracy requires a familiar viewpoint (Diamond & Carey, 1986). Much of the available research in visual expertise has studied face recognition (under the assumption that we are all *face recognition experts*) (Tarr & Gauthier, 2000), as well as experienced professionals, such as ornithologists and dog judges (Tanaka & Taylor, 1991). Furthermore, while the idea that real-world scenes are a domain of expertise akin to faces is not widespread, it is plausible (Loschky et al., 2015).

An interesting piece of evidence in favor of an expertise framework for scene perception is the important contribution afforded by low spatial frequencies for providing contextual information at the earliest stages of processing, and constraining global layout to efficiently discriminate both scenes (Oliva & Torralba, 2001) and faces (Gauthier, Curby, Skudlarski, & Epstein, 2005). Consistent with this general finding, Palmeri and Cottrell (2010) further propose a general model of expertise in which coarse information drawn from the available fine-detailed information affords the viewer the ability to create a broad continuum of related (but distinct) object classes. As long as the task type is the same (e.g., face recognition), and stimuli have a consistent structure, training a model to use low spatial frequencies seems to result in better generalization of learning to new exemplars than for models trained on higher spatial frequencies. Low spatial frequencies also produce much faster response latencies (Ivry & Robertson, 1998). Palmeri and Cottrell (2010) argue further that this type of learning is not unique to expertise. Instead, they contend that expertise is only normal learning taken to its extreme endpoint.

The second question we must answer is *how* the loss of viewpoint-dependent information affects scene categorization. A familiar viewpoint allows the viewer of a scene to make use of the neural architecture that is optimally tuned to encode scene information. This architecture is often referred to as *feed-forward*, because its layers of filters and channels are designed to produce a response within the first sweep of neural activation through the afferent visual pathways (VanRullen & Koch, 2003; VanRullen & Thorpe, 2001). However, if the information deviates from the optimal format—in this case, the upright orientation—there is a cost in terms of the utility of that information for recognizing the category of the image (Diamond & Carey, 1986; Loschky et al., 2015).

Studies investigating the time course of rapid scene categorization for aerial and terrestrial views provide evidence that during the feed-forward sweep, the chain of events from sensation to perception seems to progress at a relatively fixed rate, and that over the last third of a single eye fixation, viewers gain very little additional information. This is important given that, as defined earlier, scene gist is acquired in the first eye fixation on a scene. These conclusions were found for both aerial and terrestrial views of scenes by Loschky et al. (2015), in which images of aerial and terrestrial scenes were rapidly presented for varying amounts of processing time. This was accomplished by briefly flashing images (for 24 ms) and then, after a variable blank inter-stimulus interval (a neutral gray screen), presenting a visual mask (made from phase-randomized scene images) for 48 ms, which served to stop further information extraction from the stimulus. The processing time for the image was therefore manipulated in terms of the time from the onset of the target image to the onset of the mask, known as the target/mask *stimulus onset asynchrony* (SOA) (Breitmeyer & Ogmen, 2006; Kahneman, 1967; Turvey, 1973).

In some ways, their results showed a great deal of similarity in the visual processing of gist for both viewpoints. Viewers extracted visual information as a function of processing time at a relatively constant rate regardless of viewpoint (i.e., nearly identical accuracy×SOA slopes). Likewise, viewers reached asymptote by 211 ms SOA for both viewpoints, with essentially no improvement between 211 and 330 ms SOA for either viewpoint. Importantly, 330 ms is the typical duration of a single eye fixation when looking at scenes (Rayner, 1998), and thus, the 330 ms SOA constitutes the normal processing time limit of a single eye fixation (a defining characteristic of scene gist). Nevertheless, the results also showed clear and important differences between the gist extraction for aerial and terrestrial views. At every time point, from 24 to 330 ms SOA, the accuracy for terrestrial views was considerably higher than for aerial views, averaging 19% higher accuracy, which was 38% of the performance range. Furthermore, to attain equivalent accuracy between the two views, viewers needed triple the processing time for aerial as for terrestrial views: accuracy for 70 ms SOA for aerial views equaled accuracy for 24 ms SOA for terrestrial views, and accuracy for 211 ms for aerial views was slightly lower than accuracy for 70 ms for terrestrial views. Thus, while information seemed to be extracted automatically at a constant rate for both views, there was a lower performance ceiling for the aerial views than the terrestrial views. This suggests that reduced ability to recognize the gist of aerial views compared with terrestrial views was not due to a lack of processing time (at least within the limits of a single eye fixation) but instead, was due to inherent limits in the diagnostic information available from the aerial views compared with the terrestrial views.

There is also converging evidence that processing during the first fixation on an aerial scene is more difficult than in terrestrial scenes. Pannasch and colleagues (Pannasch, Helmert, Hansen, Larson, & Loschky, 2014) gave participants the task of categorizing scenes while their eyes were tracked. Then, for both aerial and terrestrial views, they measured the average first fixation duration, which is a common measure of visual processing difficulty (Nuthmann, Smith, Engbert, & Henderson, 2010; Rayner, 1998). They analyzed those fixations that began before image onset and ended after the onset. They then compared the durations of the portions of those fixations that were before the image onset with those after, with the prediction that the durations of the portions before image onset would not differ between views, but that the durations of the portions after onset would differ, if view had an influence on processing difficulty. The results supported the hypothesis that only the portions of the fixation durations *after* image onset differed, and they were significantly longer in the aerial view condition (239 ms) than in the terrestrial view condition (214 ms). This provides further converging evidence that aerial views are more difficult to process than terrestrial views.

These results raise the question of what would happen if one allowed viewers to examine such aerial and terrestrial images for multiple fixations. While allowing that much time to look at scenes would be outside of the realm of "scene gist recognition," it is a reasonable question. Would accuracy for aerial views still be inferior to that for terrestrial views? In the study by Pannasch et al. (2014), they allowed viewers 6.5 s (6500 ms) to do the scene categorization task, spanning roughly 20 fixations. While accuracy was understandably much higher than when viewers were given only a single eye fixation's processing time, accuracy was still significantly lower for aerial views than for terrestrial views (90.6% vs. 97.2%) (Pannasch et al., 2014), again suggesting that limited processing time is not the problem for categorizing aerial views; rather, the problem is that there is a lack of some necessary information in aerial views.

4.2.1 Quantifying Scene Structure and the Differences between Scene Views

If the visual system is provided with suboptimal information, it becomes much more expensive to process. For example, Figure 4.2a shows sample images and data from Loschky et al. (2015), in which participants were instructed to categorize aerial and terrestrial scenes that had been rotated by 0°, 45°, 90°, 135°, or 180° from their original upright orientation (0°). As shown in Figure 4.2a, terrestrial scenes showed a strong linear decrease in categorization accuracy as images were tilted away from upright. Conversely, the aerial scenes showed no effect of rotation. Interestingly, since terrestrial views are better recognized, when performance for the terrestrial views reached its low point due to rotation, it matched the performance of the aerial views. These results are consistent with the general idea that the global configuration of a scene, or its layout, as referenced to the gravitational frame is important for recognizing terrestrial views of scenes. As an illustration of this, in Figure 4.2b, the top panel shows a terrestrial view of a beach, the middle panel shows the abstract global configuration of the major regions of the beach scene (sky, water, sand), and the bottom panel shows an inverted version of the global configuration. This figure illustrates that the global configuration of a scene is greatly changed by rotation, which is consistent with the idea that the gravitational frame places important constraints on the recognition process for terrestrial views of scenes. Conversely, the effect of rotation is absent for aerial views. Therefore, when those constraints are violated by image rotation, recognition of terrestrial views suffers, whereas the same is not true for aerial views, which completely lack the constraint of

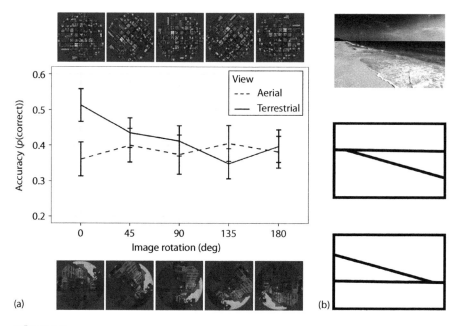

FIGURE 4.2
(a) Images and data from Experiment 2 showing a strong effect of image rotation on terrestrial scenes, but not for aerial scenes. This supports the idea that the perceptual-upright orientation of a scene is a fundamental component of normal scene perception. (b) Illustration of a dependence of a scene's global configuration on the scene's viewing orientation (e.g., changes due to image rotation). (From Loschky, L.C., et al., *J. Vis.*, 15(6:11), 1–29, 2015.)

the gravitational frame. Thus, the gravitational frame could constitute information (from an information theoretic perspective, because it provides constraints), which is present for the task of recognizing terrestrial views, but is absent for the task of recognizing aerial views.

Interestingly, in the above-mentioned Loschky et al. (2015) image rotation experiment, beyond the linear effect of rotation, there was a stronger quadratic effect, such that the poorest performance for terrestrial views was not at 180° (upside down), as one might expect, but instead, at 135° (nearly upside down and tilted obliquely to the left). The explanation for such a result was that when an image is at 180° of rotation, the perceptual upright has been completely reversed, but the orientation bias (i.e., the predominant edge orientations in the scene) has been completely preserved. For instance, in Figure 4.2b, both the upright and inverted scene configuration representations are composed of the same horizontal and diagonal edges; only the relative positions between the horizontal and diagonal edges (i.e., their configuration) have been changed. These unique biases in edge orientations are an important component for discriminating scene categories.

Prior research evaluating scene image statistics has demonstrated that natural terrestrial scenes are dominated by spatial information that is oriented horizontally, and to a lesser degree vertically, with the least information oriented obliquely (Hansen & Essock, 2004). The predominance of horizontal information is due to the horizon being ever present in natural scenes (Hansen & Essock, 2004). However, our visual systems tend to adjust in such a way that oblique orientations appear novel (Essock, DeFord, Hansen, & Sinai, 2003). Such deviations in orientation bias can be diagnostic of a scene category. For instance, a city scene will contain many vertical orientations, with skyscrapers and light-poles covering the landscape, whereas the image of mountains on the horizon will contain craggy oblique orientations that join at the peaks. In the case of the 135° rotated scenes in Figure 4.2a, not only has the scene configuration been disrupted, but so have the horizontal and vertical orientation biases. In the case of satellite images of scenes, there is no gravitational frame with which to establish an upright orientation, and thus, no absolute orientation biases are associated with particular scene categories (Loschky et al., 2015).

There is evidence from computational modeling studies suggesting that certain types of information are missing from aerial images that can be found in terrestrial images. This evidence comes from studies using versions of the Spatial Envelope (SpEn) model, which is a computational model of the visual information used to categorize (terrestrial views of) scenes (Oliva & Torralba, 2001, 2006; Torralba, 2003; Torralba & Oliva, 2002). The SpEn model does well in predicting human performance, and its underlying visual representations are well characterized (Greene & Oliva, 2009a,b; Loschky et al., 2015; Ramkumar et al., 2012). The lower levels of the SpEn model are based on well-validated assumptions about human visual processing (i.e., the processing of oriented spatial frequency contrast information in primary visual cortex), and its higher-order representations are based on human perceptual discrimination data (Oliva & Torralba, 2001, 2006; Torralba, 2003; Torralba & Oliva, 2002). Importantly, the SpEn model captures the global configuration, or "layout," of scene images, which, we have argued earlier, is very important for recognizing the gist of a scene. Thus, such computational modeling is useful in evaluating possible sources of information that may be used in human vision.

Two studies have used versions of the SpEn model to test the types of visual information available for categorizing aerial and terrestrial views of scenes (Loschky et al., 2015; Ramkumar et al., 2012). These studies took visual information from the SpEn model and used it to train different types of scene classifiers, which take patterns of features as input and return categorization decisions as output. A fundamental finding in both these studies was that terrestrial scenes were categorized more accurately than aerial scenes using two different versions of the SpEn model (Loschky et al., 2015; Ramkumar et al., 2012). This suggests that, to the extent that the SpEn captures useful information for rapidly categorizing scenes, that information is more plentiful in terrestrial views than in aerial views (i.e., something is missing from aerial views).

Furthermore, Ramkumar et al. (2012) provide a possible explanation of what is missing in terms of modeling the effect of image rotation on categorization of aerial and terrestrial scenes. They investigated the degree to which training of SpEn-based classifiers with upright scenes would generalize to testing with inverted scenes. They found that when training was on upright scenes and testing on inverted scenes, there was significantly higher categorization accuracy for aerial views than for terrestrial views. Such computational modeling results are consistent with the human behavioral results reported earlier (i.e., no effect of image rotation on accuracy of categorizing aerial views, but a significant accuracy decrement for rotated terrestrial views). This further points to global configuration being important for categorizing terrestrial scenes but not for aerial scenes—thus, global configuration may be a missing type of diagnostic information for categorizing aerial scenes.

Another key question is the degree to which the visual representations captured by the SpEn model from one view (aerial or terrestrial) generalize to categorizing images from the other view. Both studies, using slightly different versions of the SpEn model, found that when SpEn-based image classifiers were trained on aerial views and tested on terrestrial views, or vice versa, categorization accuracy dropped precipitously to only slightly above chance (Loschky et al., 2015; Ramkumar et al., 2012). Thus, the information that the SpEn model captures from one view (aerial or terrestrial) does not transfer well to the other view. Part of this lack of transfer may be due to the differential utility of global configuration information across the two views. More generally, across the two studies, the overall finding of performance drop-offs for almost all cases where training and testing differed is also consistent with expert behavior, which tends to be very viewpoint-specific. As the visual system develops, processes become more streamlined but also more view-dependent and thus, less flexible. Deviations from an expert's preferred view will, therefore, lead to a dramatic loss of efficiency.

Nevertheless, Ramkumar et al. (2012) also trained image classifiers based on human behavioral data (i.e., participants' scene categorization performance for a given view). Surprisingly, those human performance–based image classifiers showed little, if any, drop in accuracy when trained on one view and tested on the other. Namely, the patterns in the confusion matrices based on human errors tended to generalize across views much better than the errors made based on the SpEn model. However, there was slightly better generalization when training was on aerial views and testing was on terrestrial views than vice versa. Thus, the memory representations generated from aerial views generalize to terrestrial views better than the reverse. Of course, since the vast majority of humans are arguably terrestrial view experts but not aerial view experts, the poorer generalization from terrestrial views to aerial views helps explain the problem viewers have in recognizing aerial views.

This raises an important question: what is generalizable across the two views? Perhaps the simplest hypothesis is that both the information and the processes used to recognize aerial and terrestrial views are nested within one another (Shelton & Gabrieli, 2002). Namely, we can recognize both aerial and terrestrial views because they are from the same distal stimuli (i.e., objects and locations in the world), only seen from different views. Thus, one might expect that they could even share similar perceptual features. For example, an image of a city center, from either an aerial or a terrestrial view, would likely have the same hard, 90° angles regardless of view, just as a forest will likely contain a wide variety of edges meeting at different orientations regardless of view. Loschky and colleagues (2015) compared the types of errors people make when categorizing aerial scenes with those made when categorizing terrestrial scenes and found that they were moderately correlated (Spearman's $r = .63$). As shown in Figure 4.3, multidimensional scaling of confusion matrices between aerial and terrestrial scenes (Loschky et al., 2015) gives a general idea of how confusable certain scene categories are with each other and thus, how much diagnostic information overlaps between categories. The data here show a visible distinction at the superordinate category level between *natural* scenes (e.g., coast, desert, forest, mountains, and river) and most *man-made* scenes (e.g., airport, city, residential, and stadium), as the basic-level categories of scenes tended to be clustered close together within these superordinate-level categories. Interestingly, one can see in Figure 4.3 that at least one of the basic-level scene categories used in that study, Golf-Course, straddled the superordinate category boundary between natural and man-made, indicating that it was confusable at the superordinate level. Of particular interest, Golf-Course clustered somewhere between natural and man-made scenes for both aerial and terrestrial views

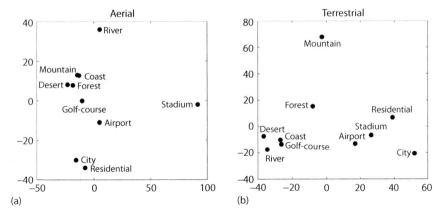

(a)   (b)

FIGURE 4.3

Multidimensional scaling of scene categorization errors between (a) aerial and (b) terrestrial views. Note how basic-level categories in both aerial and terrestrial scenes tend to cluster according to the natural/man-made distinction. (From Loschky, L.C., et al., *J. Vis.*, 15(6:11), 1–29, 2015.)

(Loschky et al., 2015). Thus, clearly something seems to be shared across both types of views—the natural/man-made distinction.

A remaining question is whether scene recognition between aerial and terrestrial scenes is due to shared visual information during early perceptual processing stages or during later semantic activation, or whether these two systems are one and the same. Perceptual processing usually progresses from coarse to fine levels of specificity (Hegde, 2008), and in the case of scene categorization, the discrimination of superordinate levels of categorization precedes basic levels of categorization (Greene & Oliva, 2009a; Loschky & Larson, 2010). If early computations recognize a terrestrial view of a scene as being natural or man-made based on distinctive visual features, it is plausible that the same visual processing and semantic networks would be coopted to perform similar computations on aerial views of scenes, and that these operations would be accurate up until a certain stage of processing. Given that we have argued that there is clearly different information available from aerial and terrestrial views of scenes (e.g., global configuration), the moderate correlation between the errors made in categorizing aerial and terrestrial scenes could be due to either (1) some shared information between both views (despite other differences in available information), (2) shared visual processing routines (Ullman, 1984), or a combination of the two. Disentangling these different possibilities is an important question for further research on aerial versus terrestrial rapid scene categorization.

4.2.2 Useful Information for Aerial Scene Categorization

A key remaining question is what sort of information is useful for categorizing aerial views. We have already discussed the importance of both global configuration and dominant scene orientation for recognizing terrestrial scene views, but not aerial views, based on our image rotation experiments. Thus, a key criterion for diagnostic information used to recognize aerial scenes is that it should survive image rotation (i.e., that it be rotation invariant). Vijayaraj and colleagues (Vijayaraj et al., 2008) carried out analyses of the image statistics of aerial and terrestrial scenes to find such candidate diagnostic visual features for aerial view scene categories. As shown in Figure 4.4, they found that the probability of the angles of co-occurring edges, specifically 90° angles, varied considerably between categories of aerial views of scenes (Vijayaraj et al., 2008, Figure 11). For example, aerial views of wooded areas have a relatively low probability of 90° angles between oriented edges (because they look like broccoli, with a predominance of curved edges), which is essentially constant as a function of distance between paired lines or edges. In contrast, aerial views of downtown areas have a higher overall probability of 90° angles (because many buildings and city blocks are rectangular), which initially linearly increase with distance between oriented edges and then reach asymptote. Such differences in the probability of the relative angles between edges as a function of distance between

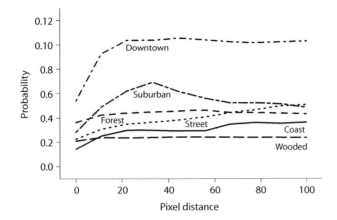

FIGURE 4.4
Probability of perpendicular (90°) pairs of co-occurring edges as a function of pixel distance between the edges, and scene categories, from aerial views (Downtown, Suburban, Coast, and Wooded) and terrestrial views (Forest and Street). (Adapted from Vijayaraj, V., et al., Overhead image statistics. Paper presented at the Applied Image Pattern Recognition Workshop, Washington, DC, 2008.)

them might be used by viewers to differentiate and categorize aerial scenes. Nevertheless, to our knowledge, this hypothesis has not been directly tested with human subjects. Importantly, such relative orientation information is rotation invariant, being determined by pairs of oriented edges having arbitrary absolute orientations, as opposed to the absolute orientations of elements relative to a fixed standard, such as the gravitational frame and the perceptual upright. Thus, the relative orientation of edges seems to be a good candidate form of information for categorizing aerial views of scenes, because it is inherently rotation-independent.

There is also reason to believe that complex textural features are useful for identifying aerial scenes. Textures can be rotation invariant (Portilla & Simoncelli, 2000), thus making them a viable candidate for providing useful information for aerial view categorization. Vijayaraj et al. (2008) analyzed the statistical power spectrum of aerial scenes at several spatial scales to determine texture's utility in discriminating aerial scenes. As shown in Figure 4.5, these data indicated that, in general, there are apparent differences between scene categories in terms of the correlation of paired pixel luminance values as a function of pixel distance. In other words, to capture a texture that would be repeated across a homogeneous landscape (e.g., forests, deserts, etc.), one would need only a small sample of the image, because the textures of one section selected at random would likely match another selected at random. Scenes whose textures are more heterogeneous (e.g., a commercial complex) would need a much larger area to replicate texture patterns across neighboring sections, as seen in Figure 4.5. Note that texture information often includes specific angles of co-occurring edges

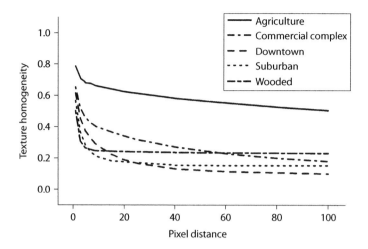

FIGURE 4.5

Texture homogeneity as a function of pixel offset distance. Notice the shallower slopes for scenes with smooth textures such as "Agriculture" and "Commercial Complex," while rough-textured scenes such as "Wooded" and "Suburban" show sharper declines in texture homogeneity as a function of pixel distance in aerial views. (Adapted from Vijayaraj, V., et al., Overhead image statistics. Paper presented at the Applied Image Pattern Recognition Workshop, Washington, DC, 2008.)

(e.g., 90° angles), so both may provide diagnostic aerial scene information together. Other researchers have also used rotation-invariant texture descriptors for identifying specific categories of scenes, such as man-made built-up areas, surrounded by natural landscape elements such as forests, in aerial views of scenes (Gamba, Pesaresi, Molch, Gerhardinger, & Lisini, 2008; Pesaresi & Gerhardinger, 2011). These have used the gray-level co-occurrence matrix, which is a rotation-invariant texture descriptor (Gamba et al., 2008; Pesaresi & Gerhardinger, 2011).

   Loschky and colleagues (2015) tested the utility of texture information for rapidly categorizing scenes by using a texture analysis and synthesis algorithm (Portilla & Simoncelli, 2000) to create synthesized texture versions of aerial and terrestrial scenes. Images (aerial vs. terrestrial; original vs. synthesized texture) were flashed for either short (35 ms SOA) or long (330 ms SOA) processing times. The results showed that participants were far more accurate at categorizing normal images than texture images. However, when participants were given longer processing times, the aerial texture scenes showed a slight but significant increase in categorization accuracy (from 10% [chance] to 20% [double chance]) compared with the terrestrial scenes (which remained at chance). Nevertheless, the aerial textures were accurately categorized only 20% of the time with the longest processing times, whereas normal aerial scenes were accurately categorized approximately 60% of the time. Furthermore, the correlations between confusion matrices for normal and texture versions of scenes were low and comparable for both aerial scenes ($r = .298$) and terrestrial scenes ($r = 0.270$), suggesting that texture, as synthesized by the Portilla and Simoncelli model (2000), plays little role in rapid scene categorization for either view.

There are several limitations to these results, however. First, the Portilla and Simoncelli (2000) texture algorithm may not be an accurate representation of texture filters that exist in the human visual cortex (Wallis, Bethge, & Wichmann, 2016). For example, it is more likely that texture-like processing by the visual system differs between central and peripheral vision rather than being uniform throughout the visual field (Ehinger & Rosenholtz, 2016; Freeman & Simoncelli, 2011). Second, the synthesized texture images contained global textures having relatively homogeneously repeated patterns, which would disrupt the global heterogeneous configurations that are typical of some aerial scene categories. For example, Figure 4.6 shows that a major difference between aerial views of coasts versus deserts is that beaches have a boundary between water and sand (or land), whereas deserts generally do not. This unique global configuration (a boundary between water and land) in an aerial view of a coast is likely to be lost when using the Portilla and Simoncelli (2000) algorithm to synthesize a texture version of an aerial coast scene, because the algorithm is targeted at analyzing and synthesizing homogeneous repeated patterns (e.g., artificial/periodic textures such as checker boards or photographic/structured textures such as a bin of peanuts). Thus, aerial views of coasts versus deserts are far less

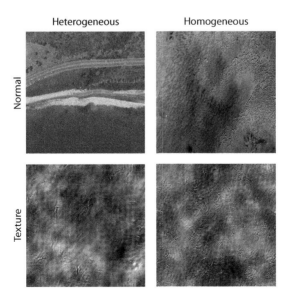

FIGURE 4.6
Sample aerial images of a coast and desert (above) with their texture-synthesized counterparts (below). Note that while the texture-synthesized versions are quite similar, the original images are distinguishable by their configuration, suggesting that local, segmented texture may be important for preserving the unique properties of a scene.

distinguishable when presented as synthesized texture image versions than the original image versions, because the original images contain such distinctive global configurations. Despite the differences in luminance between sand and water, the characteristic boundary will be lost, making beach and desert textures indistinguishable.

It is possible that texture information that maintains the global configuration of a scene (e.g., the boundary between water and land), such as the modifications of the Portilla and Simoncelli (2000) texture model by Freeman and Simoncelli (2011) or Rosenholtz and colleagues (Ehinger & Rosenholtz, 2016; Rosenholtz, Huang, Raj, Balas, & Ilie, 2012), may be of greater utility in recognizing not only terrestrial views but also aerial views of scenes. Nevertheless, we would not expect to find that rotating such texture images would have any effect on viewers' ability to categorize them. We should also note the important difference between a computational model that analyzes and describes a texture and one that synthesizes a texture. Since the primary emphasis of this chapter is on understanding how aerial views are recognized and categorized, only texture analysis is necessary. However, to empirically test the utility of texture information in recognizing scenes using human data, it is convenient to be able to synthesize texture versions of aerial (and terrestrial) scenes.

4.3 Satellite Imagery: A Biological Perspective

Despite the lower efficiency and accuracy of gist recognition for aerial views compared with terrestrial views of scenes, people are still quite good at categorizing aerial scenes. Across several studies investigating the human categorization of aerial views of natural scenes, accuracy has been shown to be well above chance (Lloyd, Hodgson, & Stokes, 2002; Loschky et al., 2015; Ramkumar et al., 2012). Furthermore, when viewing aerial views of scenes in a visual search task, if people are told to look for a target in a specified region (e.g., water, fields, roads, buildings, or foliage), they can use this information to guide their visual attention within the first three to five eye fixations (i.e., ~900–1500 ms) depending on the region category (buildings and water being fastest to guide attention, and roads being slowest) (Zelinsky & Schmidt, 2009).

Perhaps an unstated, but critical, difference between aerial and terrestrial views is in terms of the behaviors they afford to the viewer (Gibson, 1979). In fact, a large body of comparative (i.e., cross-species) cognition literature has been devoted to understanding how species-specific cognitive modules (e.g., *behavior systems*) are activated by task-specific stimuli to elicit the appropriate behavior (Hogan, 1994). For instance, a terrestrial view may be more likely to activate behavior systems that are relevant to directly accessing and interacting with one's environment, because that is what our prior experiences have conditioned us to expect. Conversely, most people (except pilots) lack experience of interacting with their environment from aerial views of scenes. However, people's common uses of aerial views imply that they provide us with a global understanding of sections of land and are more useful for planning routes and understanding spatial relationships between landmarks so that we can navigate more efficiently.

The differential spatial representations useful for traversing terrestrial and aerial scenes are often referred to as *route* and *survey* information, respectively (Siegel & White, 1975; Tversky, 1991). Route information is generally regarded as being encoded serially, as objects occlude the landscape and limit the understanding of exactly where all objects might exist from a given view of the space. Survey information is akin to looking at a map and being able to plan and anticipate changes in paths or the presence of certain landmarks to understand one's relative location. As one accumulates route information, one can then develop a cognitive map rooted in survey information (Tversky, 1991, 1993). Likewise, there are numerous shared neural substrates important for the encoding of route and survey information. For instance, both views seem to recruit brain regions from dorsal (*where*) and ventral (*what*) pathways as well as the fusiform gyrus (Shelton & Gabrieli, 2002), which is critical for configural/holistic processing (Gauthier, Skudlarski, Gore, & Anderson, 2000; Gauthier & Tarr, 2002). However, several of these shared substrates showed different levels of activation between route and survey perspectives. For instance, both hemispheres of the superior parietal cortex and the medial temporal lobes (MT) showed greater activation for route information, suggesting a stronger connection between spatial and object-based perception and encoding for route learning (Shelton & Gabrieli, 2002).

Additionally, several brain regions heavily implicated in scene perception were activated by route learning but not by survey learning (e.g., the parahippocampal place area and the precuneus) (Maguire, Frackiowiak, & Frith, 1997; Shelton & Gabrieli, 2002). Conversely, Shelton and Gabrieli (2002) found scant evidence for regions of the brain that were devoted purely to survey information and instead, found stronger activation for survey encoding in those shared regions. The above neurophysiological evidence suggests that route information, which is associated with terrestrial views, is more fundamental to human vision than is survey information, which is associated with an aerial perspective. This survey information had to be laboriously inferred by cartographers prior to the development of aerial photography.

While we can encode and use information from the aerial (survey) perspective, its practical use is still rooted in the need to apply that information within a terrestrial (route) perspective. Furthermore, generating one's mental map of the world is not just limited to isolated landmarks but may include rich networks of spatially and temporally contiguous episodic memories from navigational events as well (Tversky, 1993). Thus, egocentric route information provides an instance of viewpoint-dependent visual information. Conversely, allocentric survey information provides a broad map of visual space and is view-independent. A critical component of expertise exists in the structure of the stimulus. Specifically, expert-level visual categorization performance is well known to be view-dependent (Diamond & Carey, 1986; Gauthier & Tarr, 1997, 2002). Thus, in addition to aerial views lacking a canonical viewing orientation, the very nature of aerial scenes is incompatible with our view-based expertise for terrestrial scenes and in terms of how we habitually interact with terrestrial scenes from a route perspective. Thus, all of these factors pose serious hurdles for developing aerial scene expertise.

What do we know about the development of expertise in recognizing aerial views of scenes? First, there is evidence that extensive experience with aerial scenes can lead to improvements in processing efficiency. Lloyd et al. (2002) found that geographers were significantly more accurate, faster, and more confident than non-geographers at categorizing the land use of aerial views of scenes. Note, however, that in that study, "geographers" included geography majors, graduate students, and faculty members and thus had tremendous variability in their levels of expertise. A more recent study of recognition memory for aerial scenes (Sikl & Svatonova, 2015) had more differentiated levels of expertise, between experts (who had on average > 15 years of experience and viewed aerial scenes daily for their work), untrained first-year geography students, and untrained first-year psychology students. They showed that experts had significantly more accurate recognition memory for aerial views of scenes but responded significantly more slowly than either group of non-experts (perhaps due to the experts being, on average, 20 years older than the non-experts and processing speed decreasing with age). Second, such expertise is not something that can be gained in a semester or a year. Sikl and Svatonova found no differences in the recognition memory accuracy or speed for aerial views of scenes between first-year geography students and first-year psychology students. Likewise, an unpublished study by the authors of this chapter found that there was no difference in the rapid categorization accuracy of aerial views of scenes between students in an introductory geographical information systems course and students in an art photography course. Clearly, there is a need for more studies to investigate the effects of expertise on aerial view perception.

For any psychological phenomenon, we generally find that the driving forces of evolution and experience work in tandem. However, one can argue that in the development of the nested visual systems used to perceive not only terrestrial views of scenes but also aerial views, biology should play a larger role than experience. From an evolutionary biological perspective, learning through experience is presumed to be beneficial for an organism only when environmental information is consistent within generations but varies across generations. If information in the environment is constant across generations, then processing modules tend to be more hard-wired rather than amenable to learning (Stephens, 1991). In the case of route versus survey representations, before the advent of airplanes and satellites or the evolutionarily somewhat less recent advent of map-making, eons of evolutionary history would dictate that survey learning could only be a consequence of compiled route information. Thus, there should be an inherent advantage to our brain's ability to efficiently encode information from an earthbound perspective. Despite this proposed evolutionarily based perceptual and cognitive limitation, our human ability to form allocentric, survey knowledge–based cognitive maps, which eventually enabled our species to create physical maps, is perhaps one reason why aerial scene gist is not so foreign that we are functionally incapable of recognizing the earth from such a view. That is, our ability to think in terms of the allocentric representations we use to create maps has been shown to be a valid representation of terrestrial space by the similarities between maps and (at least some aspects of) satellite photography. This cognitive ability to construct, or at least correctly interpret, maps would seem important in learning to recognize aerial views of scenes.

Birds, on the other hand, have an entirely different evolutionary history. With the exception of some ground-nesting species, birds were offered a very different visual perspective from early in life. In one study (Kirkpatrick, Sears, Hansen, & Loschky, 2014), lab-raised pigeons (with no flight experience) were trained to categorize images of coasts and mountains. The birds were trained to peck a screen when an image from the target category was presented, with each pigeon trained to respond only to one of the two categories, only coasts or only mountains, for the duration of their training and testing. The authors then determined how easily the pigeons learned to differentiate these two scene categories. Importantly, the images also differed between those photographed from three views relative to the ground plane: aerial (90°), so-called *bird's eye* (45°), or terrestrial (0°). This enabled testing whether the pigeon's visual system is optimized for categorizing images from a particular view. The results demonstrated that the pigeons more rapidly learned to categorize the scenes photographed from the aerial and bird's eye views than from the terrestrial views. This is consistent with an evolutionary argument that humans, who evolved to recognize terrestrial views but not aerial views, are better at recognizing the former than the latter, while pigeons evolved to most easily learn to recognize bird's eye and aerial views. However, this conclusion is weakened by two important caveats. First, a follow-up analysis of the image statistics of the three views showed that—at a statistical level—the bird's eye and aerial views were more discriminable than the terrestrial scenes, and this significantly predicted the pigeons' image discrimination performance. Thus, if people were presented with the same stimuli, one might find the same results (i.e., aerial and bird's eye views producing better performance). Second, the number of image categories used in this study was quite limited, due to the constraints of the pigeon's intellect and the number of pigeons that could be housed and trained.

Indeed, the use of only these two scene categories by Kirkpatrick et al. (2014) may explain why their image statistical analyses suggested that aerial views were more discriminable than terrestrial views, while similar image statistical analyses in other studies using six scene categories (Ramkumar, Hansen, Pannasch, & Loschky, 2016) or 10 categories (Loschky et al., 2015) found that terrestrial views were more discriminable than aerial views (as discussed in detail earlier). Furthermore, human categorization data from Experiment 1 of Loschky et al. (2015) demonstrated that "mountain" and "coast" scenes were recognized with a much higher degree of accuracy from the terrestrial view, and that these two categories were far from being the most accurately recognized among aerial scenes. Therefore, the roles of evolutionary constraints versus experience in the ability to rapidly categorize aerial versus terrestrial views require further investigation.

4.4 Summary

Throughout this chapter, we have discussed the limitations that we as human beings experience when viewing the earth from an aerial perspective. Our review of the small literature on this topic has shown that there seem to be overlapping visual processing routines, and possibly some overlapping visual features, between aerial and terrestrial views, which facilitate the recognition of aerial views. However, our review has also shown several core differences that exist between the available visual information from aerial and terrestrial views and possibly, between the visual routines used to rapidly categorize the two views. A key difference appears to be that the global configurations of aerial views are far more variable than terrestrial views of scenes, because terrestrial views (but not aerial views) are constrained by the gravitational frame to (normally) being seen aligned with the perceptual upright. Without this important perceptual constant, our visual system—which has evolved within an earthbound existence—is far less efficient. Although our neural architecture may not be optimized for the perception of satellite imagery, humans have developed the ability to create allocentric survey mental representations, which appear to capture some important spatial aspects of what an aerial view represents. This raises interesting questions for further research on training people to more efficiently recognize and interact with satellite photography. Would standardizing the views, orientations, or scales of satellite images, to provide learners with more consistent perceptual input, facilitate their learning to recognize them? Or, would allowing learners to view the same scenes from multiple views aid in the early stages of learning to recognize aerial views?

References

Biederman, I., Rabinowitz, J., Glass, A., & Stacy, E. (1974). On the information extracted from a glance at a scene. *Journal of Experimental Psychology*, 103, 597–600.

Breitmeyer, B. G., & Ogmen, H. (2006). *Visual Masking: Time Slices through Conscious and Unconscious Vision*. New York: Oxford University Press.

Diamond, R., & Carey, S. (1986). Why faces are and are not special: An effect of expertise. *Journal of Experimental Psychology: General*, 115(2), 107–117.

Ehinger, K. A., & Rosenholtz, R. (2016). A general account of peripheral encoding also predicts scene perception performance. *Journal of Vision*, 16(2):13, 1–19.

Essock, E. A., DeFord, J. K., Hansen, B. C., & Sinai, M. J. (2003). Oblique stimuli are seen best (not worst!) in naturalistic broad-band stimuli: A horizontal effect. *Vision Research*, 43(12), 1329–1935.

Fei-Fei, L., Iyer, A., Koch, C., & Perona, P. (2007). What do we perceive in a glance of a real-world scene? *Journal of Vision*, 7(1:10), 1–29.

Freeman, J., & Simoncelli, E. P. (2011). Metamers of the ventral stream. *Nature Neuroscience*, 14, 1195–1201.

Gamba, P., Pesaresi, M., Molch, K., Gerhardinger, A., & Lisini, G. (2008, July 7–11). Anisotropic rotation invariant built-up presence index: Applications to SAR data. Paper presented at the IGARSS 2008–2008 IEEE International Geoscience and Remote Sensing Symposium, Boston, MA.

Gauthier, I., Curby, K. M., Skudlarski, P., & Epstein, R. A. (2005). Individual differences in FFA activity suggest independent processing at different spatial scales. *Cognitive, Affective, and Behavioral Neuroscience*, 5(2), 222–234.

Gauthier, I., Skudlarski, P., Gore, J. C., & Anderson, A. W. (2000). Expertise for cars and birds recruits brain involved in face recognition. *Nature Neuroscience*, 3(2), 191–197.

Gauthier, I., & Tarr, M. J. (1997). Becoming a "Greeble" expert: Exploring mechanisms for face recognition. *Vision Research*, 37(12), 1673–1682.

Gauthier, I., & Tarr, M. J. (2002). Unraveling the mechanisms for expert object recognition: Bridging brain activity and behavior. *Journal of Experimental Psychology: Human Perception and Performance*, 28(2), 431–446.

Gibson, J. J. (1979). *The Ecological Approach to Visual Perception: Classic Edition*. New York: Psychology Press.

Greene, M. R., & Oliva, A. (2009a). The briefest of glances: The time course of natural scene understanding. *Psychological Science*, 20(4), 464–472.

Greene, M. R., & Oliva, A. (2009b). Recognition of natural scenes from global properties: Seeing the forest without representing the trees. *Cognitive Psychology*, 58(2), 137–176.

Haji-Khamneh, B., & Harris, L. R. (2010). How different types of scenes affect the subjective visual vertical (SVV) and the perceptual upright (PU). *Vision Research*, 50(17), 1720–1727.

Hansen, B. C., & Essock, E. A. (2004). A horizontal bias in human visual processing of orientation and its correspondence to the structural components of natural scenes. *Journal of Vision*, 4(12), 1044–1060.

Harris, L. R., Jenkin, M., Dyde, R. T., & Jenkin, H. (2011). Enhancing visual cues to orientation: Suggestions for space travelers and the elderly. In A. M. Green, C. E. Chapman, J. F. Kalaska & F. Lepore (Eds.), *Progress in Brain Research* (Vol. 191, pp. 133–142). Oxford: Elsevier.

Hegde, J. (2008). Time course of visual perception: Coarse-to-fine processing and beyond. *Progress in Neurobiology*, 84(4), 405–439.

Hogan, J. A. (1994). Structure and development of behavior systems. *Psychonomic Bulletin and Review*, 1(4), 439–450.

Ivry, R., & Robertson, L. C. (1998). *The Two Sides of Perception*. Cambridge, MA: MIT Press.

Kahneman, D. (1967). An onset-onset law for one case of apparent motion and meta-contrast. *Perception and Psychophysics*, 2(12-A), 577–584.

Kirkpatrick, K., Sears, T., Hansen, B. C., & Loschky, L. C. (2014). Scene gist categorization by pigeons. *Journal of Experimental Psychology: Animal Behavioral Processes*, 40(2), 162–177.

Larson, A. M., Freeman, T. E., Ringer, R. V., & Loschky, L. C. (2014). The spatiotemporal dynamics of scene gist recognition. *Journal of Experimental Psychology: Human Perception and Performance*, 40(2), 471–487.

Lloyd, R., Hodgson, M. E., & Stokes, A. (2002). Visual categorization with aerial photographs. *Annals of the Association of American Geographers*, 92(2), 241–266.

Loschky, L. C., Hansen, B. C., Sethi, A., & Pydimari, T. (2010). The role of higher-order image statistics in masking scene gist recognition. *Attention, Perception and Psychophysics*, 72(2), 427–444.

Loschky, L. C., & Larson, A. M. (2008). Localized information is necessary for scene categorization, including the natural/man-made distinction. *Journal of Vision*, 8(1), 4, 1–9.

Loschky, L. C., & Larson, A. M. (2010). The natural/man-made distinction is made prior to basic-level distinctions in scene gist processing. *Visual Cognition*, 18(4), 513–536.

Loschky, L. C., Ringer, R., Ellis, K., & Hansen, B. C. (2015). Comparing rapid scene categorization of aerial and terrestrial views: A new perspective on scene gist. *Journal of Vision*, 15(6:11), 1–29.

Maguire, E. A., Frackowiak, R. S. J., & Frith, C. J. (1997). Recalling routes around London: Activation of the right hippocampus in taxi drivers. *Journal of Neuroscience*, 17(18), 7103–7110.

Nuthmann, A., Smith, T. J., Engbert, R., & Henderson, J. M. (2010). CRISP: A computational model of fixation durations in scene viewing. *Psychological Review*, 117(2), 382–405.

Oliva, A. (2005). Gist of a scene. In L. Itti, G. Rees & J. K. Tsotsos (Eds.), *Neurobiology of Attention* (pp. 251–256). Burlington, MA: Elsevier Academic.

Oliva, A., & Torralba, A. (2001). Modeling the shape of the scene: A holistic representation of the spatial envelope. *International Journal of Computer Vision*, 42(3), 145–175.

Oliva, A., & Torralba, A. (2006). Building the gist of a scene: The role of global image features in recognition. *Progress in Brain Research, Special Issue on Visual Perception*, 155, 23–36.

Palmeri, T. J., & Cottrell, G. W. (2010). Modeling perceptual expertise. In I. A. Gauthier, M. J. Tarr & D. Bub (Eds.), *Perceptual Expertise: Bridging Brain and Behavior* (pp. 197–244). New York: Oxford University Press.

Pannasch, S., Helmert, J. R., Hansen, B. C., Larson, A. M., & Loschky, L. C. (2014). Commonalities and differences in eye movement behavior when exploring aerial and terrestrial scenes. In M. Buchroithner, N. Prechtel & D. Burghardt (Eds.), *Cartography from Pole to Pole* (pp. 421–430). Berlin: Springer.

Pesaresi, M., & Gerhardinger, A. (2011). Improved textural built-up presence index for automatic recognition of human settlements in arid regions with scattered vegetation. *IEEE Journal of Selected Topics in Applied Earth Observations and Remote Sensing*, 4(1), 16–26.

Portilla, J., & Simoncelli, E. P. (2000). A parametric texture model based on joint statistics of complex wavelet coefficients. *International Journal of Computer Vision*, 40(1), 49–71.

Potter, M. C., & Levy, E. I. (1969). Recognition memory for a rapid sequence of pictures. *Journal of Experimental Psychology*, 81(1), 10–15.

Ramkumar, P., Hansen, B. C., Pannasch, S., & Loschky, L. C. (2016). Visual information representation and rapid-scene categorization are simultaneous across cortex: An MEG study. *Neuroimage*, 134, 295–304.

Ramkumar, P., Pannasch, S., Hansen, B. C., Larson, A. M., & Loschky, L. C. (2012). How does the brain represent visual scenes? A neuromagnetic scene categorization study. In G. Langs, I. Rish, M. Grosse-Wentrup & B. Murphy (Eds.), *Machine Learning and Interpretation in Neuroimaging* (Vol. 7263, pp. 93–100). Berlin: Springer.

Rayner, K. (1998). Eye movements in reading and information processing: 20 years of research. *Psychological Bulletin*, 124(3), 372–422.

Rosenholtz, R., Huang, J., Raj, A., Balas, B. J., & Ilie, L. (2012). A summary statistic representation in peripheral vision explains visual search. *Journal of Vision*, 12(4), 14.

Schyns, P. G., & Oliva, A. (1994). From blobs to boundary edges: Evidence for time- and spatial-scale-dependent scene recognition. *Psychological Science*, 5, 195–200.

Shelton, A. L., & Gabrieli, J. D. E. (2002). Neural correlates of encoding space from route and survey perspectives. *Journal of Neuroscience*, 22(7), 2711–2717.

Siegel, A. W., & White, S. H. (1975). The development of spatial representations of large-scale environments. *Advances in Child Development and Behavior*, 10, 9–55.

Sikl, R., & Svatonova, H. (2015). Visual recognition memory for aerial photographs [Abstract]. *Perception*, 44, 9.

Stephens, D. W. (1991). Change, regularity, and value in the evolution of animal learning. *Behavioral Ecology*, 2(1), 77–89.

Tanaka, J. W., & Taylor, M. J. (1991). Object categories and expertise: Is the basic level in the eye of the beholder? *Cognitive Psychology*, 23(3), 457–482.

Tarr, M. J., & Gauthier, I. (2000). FFA: A flexible fusiform area for subordinate-level visual processing automatized by expertise. *Nature Neuroscience*, 3(8), 764–769.

Thorpe, S. J., Fize, D., & Marlot, C. (1996). Speed of processing in the human visual system. *Nature*, 381(6582), 520–522.

Torralba, A. (2003). Modeling global scene factors in attention. *Journal of the Optical Society of America, A, Optics, Image Science and Vision*, 20(7), 1407–1418.

Torralba, A., & Oliva, A. (2002). Depth estimation from image structure. *IEEE Transactions on Pattern Analysis and Machine Intelligence*, 24(9), 1226–1238.

Turvey, M. T. (1973). On peripheral and central processes in vision: Inferences from an information-processing analysis of masking with patterned stimuli. *Psychological Review*, 80(1), 1–52.

Tversky, B. (1991). Spatial mental models. *Psychology of Learning and Motivation*, 27, 109–145.

Tversky, B. (1993). Cognitive maps, cognitive collages, and spatial mental models. In A. U. Frank & I. Campari (Eds.), *Spatial Information Theory: A Theoretical Basis for GIS*, (pp. 14–24). Berlin: Springer.

Ullman, S. (1984). Visual routines. *Cognition*, 18(1–3), 97–159.

VanRullen, R., & Koch, C. (2003). Visual selective behavior can be triggered by a feed-forward process. *Journal of Cognitive Neuroscience*, 15(2), 209–217.

VanRullen, R., & Thorpe, S. J. (2001). The time course of visual processing: From early perception to decision-making. *Journal of Cognitive Neuroscience*, 13(4), 454–461.

Vijayaraj, V., Cheriyadat, A. M., Sallee, P., Colder, B., Vatsavai, R. R., Bright, E. A., & Bhaduri, B. L. (2008). Overhead image statistics. Paper presented at the Applied Image Pattern Recognition Workshop, Washington, DC.

Wallis, T. S. A., Bethge, M., & Wichmann, F. A. (2016). Testing models of peripheral encoding using metamerism in an oddity paradigm. *Journal of Vision*, 16(2), 4: 1–30.

Zelinsky, G. J., & Schmidt, J. (2009). An effect of referential scene constraint on search implies scene segmentation. *Visual Cognition*, 17(6–7), 1004–1028.

5

Eye-Tracking Evaluation of Non-Photorealistic Maps of Cities and Photo-Realistic Visualization of an Extinct Village

Stanislav Popelka

CONTENTS

KEYWORDS: *cartography, photorealism, evaluation, eye-tracking, 3D model*

5.1 Introduction

The chapter describes an eye-tracking study focused on the evaluation of two-dimensional (2D) and pseudo-three-dimensional (3D) visualization of urban areas in cartography. 3D visualization is used by an increasing number of visualization applications and work systems. However, there is still little knowledge about when it is appropriate to use 3D visualizations and how 3D can be used in visualization most efficiently. Few studies have focused on the evaluation of 3D in maps. Most use the questionnaire as the main investigation method. Two studies described in this chapter were performed with the use of eye-tracking, which allows the researcher to analyze user strategy during task completion and not only users' final judgments or decisions.

The chapter is focused on the evaluation of photorealistic and non-photorealistic 3D visualization of urban areas via two eye-tracking experiments. The first deals with non-photorealistic visualization of cities in map portals and the second with images of a created 3D model of an extinct village. The goal of the chapter is to determine the efficiency of 3D visualization of urban areas. The chapter consists of two case studies involving a non-photorealistic and a photorealistic visualization.

5.2 3D Geovisualization

There are several perspectives on the term *3D* in cartography. In one of the first articles addressing 3D maps, Menno-Jan Kraak (1988) stated that an image would be considered 3D if it contained those stimuli or depth cues that make it possible to perceive something as 3D.

Wood, Kirschenbauer, Döllner, Lopes, and Bodum (2005) described the role of three-dimensionality used in processes of both visualization and representation of 3D objects and space. To clarify the term *3D*, they considered the model of a general visualization pipeline. It is possible to distinguish five levels of dimensionality, which correspond to the various stages within a visualization process: data management, data assembly, visual mapping, rendering, and display (Haber & McNabb, 1990; Upson et al., 1989). The term *3D* is associated with the phase of "visual representations of data," but it can also be part of the outputs from all other stages of the process (Wood et al., 2005).

According to Bleisch (2012), *3D geovisualization* is quite a generic term, which is applied to a range of 3D visualizations representing the real world, parts of the real world, or other data having a spatial reference. Especially with the advent of virtual globes or geobrowsers such as Google Earth, 3D visualizations are increasingly popular, and many people know about them, even though they may not refer to them as 3D. Studies described in this chapter use pseudo-3D visualization (or perspective images), but they are referred to as 3D.

According to Döllner (2008), the two basic forms of 3D cartography are non-photorealistic and photorealistic visualizations. In terms of computer graphics, photorealistic and non-photorealistic visualization primarily differ in the way of shape representation, coloring, lighting, shading, and shadowing.

5.2.1 Non-Photorealistic Visualization

Since about 1990, non-photorealism has established itself as a key sub-discipline in computer graphics. The term *non-photorealistic computer graphics* denotes depictions that reflect true or imaginary scenes in stylistic, illustrative, or artistic styles. The general characteristics of non-photorealistic rendering techniques include the ability to sketch geometric objects and scenes, to reduce the visual complexity of images, as well as to imitate and extend classical depiction techniques known from scientific and cartographic illustrations (Döllner & Buchholz, 2005).

Durand (2002) stated that non-photorealistic computer graphics techniques offer extensive control over expressivity, clarity, and aesthetics; thereby, the resulting pictures "can be more effective at conveying information, more expressive or more beautiful" (p. 1). Jedlicka, Cerba, & Hajek (2013) suggest that the techniques have limitations (generalization, simplification, non-perspective projections, distortions, etc.). Limitations such as distortion might lead to slower reading or incorrect interpretation of the map (Popelka & Dolezalova, 2016).

These general facts also apply for the visualization of urban areas. In cartography, the visual representation of city models has a long history and has yielded many principles for drawings of this category. The most prominent examples include panoramic maps of cities and landscapes as well as bird's-eye views of cities (Döllner, 2007). Internet map portals use non-photorealistic 3D models at different levels of abstraction as an enhancement of the map, especially for large cities.

These technical capabilities notwithstanding, cartographic theory and principles for 3D map design are almost non-existent (Haeberling, 2002). To create informative and sophisticated 3D maps, design guidelines should be formulated based on cartographic principles and research. The aim of these guidelines is to give cartographers additional stimuli to generate informative and useful representations, ultimately benefiting the map user.

5.2.2 Photorealistic Visualization

Photorealistic visualization of urban areas is prevalent in three main areas: visualization of existing cities, cultural heritage documentation, and virtual reconstruction of already non-existing sites.

Photorealistic 3D visualization of existing cities became popular due to the ubiquity of Google Earth and similar applications, whereby models of large areas are created with automatic or semiautomatic acquisition methods. Models are created in great detail, with aerial photos used as base maps. However, for small-scale maps depicting a large area, it is problematic to use photorealistic visualization, since the details of 3D buildings are difficult to discern.

Photorealistic 3D visualization is also commonly used for cultural and natural heritage documentation, because heritage sites (natural, cultural, or mixed) suffer from wars, natural disasters, weather changes, and human negligence. The importance of cultural heritage documentation is well recognized, and there is increasing pressure to document and preserve the heritage digitally (Remondino & Rizzi, 2010). For example, 3D visualization for cultural heritage promotion was used by Koutsoudis, Arnaoutoglou, and Chamzas (2007), who used an open-source system for realistic virtual walkthroughs of the old city of Xanthi in Greece.

The third main application of 3D visualization of urban areas is its use to create a virtual reconstruction of extinct areas. This approach is widely used for virtual reconstructions of parts of the city or individual buildings (Cartwright, 2006; Remondino et al., 2009). An example of photorealistic visualization of an entire city is a 3D model of Rome in 320 BC (Guidi, Frischer, & Lucenti, 2007). Individual buildings were modeled in great detail, although for most of the buildings, it is not clear what the source of information about the structures was, or what the structures looked like.

The creation of photorealistic 3D visualization is a cost- and time-demanding process. For some fields of use, it is not confirmed whether the photorealism is necessary or beneficial for the users. Usability evaluation of the photorealistic 3D visualization is a way that can help to clarify this issue.

5.2.3 Evaluation of 3D

While 3D visualization is used in an increasing number of applications, little is known about how 3D can be most efficiently used in visualization. For many aspects involved in 3D geovisualization, the theory and design guidelines do not yet exist, and suitable evaluation is needed (Slocum et al., 2001). According to Montello (2009), the study of cognition (evaluation tests) has been a fundamental research domain for GIScience from its beginning, and this status continues to this day. Nevertheless, as Konečný, Kubíček, Stachoň, and Šašinka (2011) highlight, the creation of the usability tests for different types of maps and visualizations is quite a challenge.

Usability studies of 3D geovisualization are often focused on the usability of technical aspects of a visualization (e.g., how usable is a specific application; is the user able to orientate himself; are the buttons in places where they are accessible?) (Bleisch, 2012). Such analyses can be valuable but not always sufficient (Greenberg & Buxton, 2008). Aspects such as the fulfillment of purpose or the appropriateness of a 3D geovisualization are more difficult to measure and more rarely attempted (Lam, Bertini, Isenberg, Plaisant, & Carpendale, 2012). John, Cowen, Smallman, and Oonk (2001) performed six experiments with 3D perspective views using stimuli from air traffic control and military command areas. Perspective views were superior to 2D maps in three experiments. In the other three, 2D was superior to perspective views.

Few studies have used the questionnaire as the main investigation method for 3D visualization usability studies. Savage, Wiebe, and Devine (2004) and Petrovič and Mašera (2004) analyzed users' preferences for 2D and 3D maps. Schobesberger and Patterson (2007) investigated differences between a 2D and a 3D map of the Zion National Park in Utah. Haeberling (2002) evaluated design variables for 3D maps.

In a few studies, eye-tracking has been used for the evaluation of 3D maps. Putto, Kettunen, Torniainen, Krause, & Tiina Sarjakoski (2014) investigated the perception of elevation information on maps. The effect of three different visualizations of elevation information on eye movements and performance was investigated in visual search, area selection, and route planning tasks. The results of the study showed that the relief shading did not slow down the performance in terms of either response time or eye-movement measures. Interaction with a 3D geo-browser under time pressure was evaluated by Wilkening and Fabrikant (2013). The study does not provide any empirical evidence for the added value of using 3D interactive map displays to solve 3D tasks. The perception of 2D and 3D terrain maps was investigated in Popelka and Brychtova (2013). Two eye-tracking tests were used for observing user perception of a pair of terrain maps. On one map, the terrain was represented by contour lines, whereas the second map contained a perspectival view of the same data (Figure 5.1). Analysis of the questionnaire results has shown that the majority of participants prefer perspectival (pseudo-3D) visualization, but statistical analysis of recorded data did not find significant differences between user perceptions of both map types.

Maps of urban areas were studied with the use of eye-tracking in a study by Simone, Presta, and Protti (2014), who analyzed urban change maps. Fuhrmann, Komogortsev, and Tamir (2009) analyzed differences between

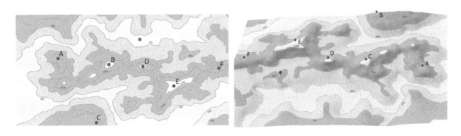

FIGURE 5.1
An example of stimuli from the study of Popelka and Brychtova. (From Popelka, S. and Brychtova, A., *Cartogr. J.*, 50(3), 240–246, 2013.)

perception of a 2D map of a city and its holographic equivalent with the use of eye-tracking. The results showed a higher time on task and traversed eye-path (scanpath length) for the 3D hologram. The authors stated that these results might not necessarily indicate lower effectiveness in route planning, because in the case of the hologram, they cannot rule out an "awe" effect caused by the novelty of this 3D visualization. A mobile eye-tracker was used in the studies of Kiefer, Giannopoulos, and Raubal (2014) and Kiefer, Straub, and Raubal (2012), who analyzed wayfinding behavior in a city. The gaze data collected during the experiments showed that successful participants paid significantly more visual attention to those symbols on the map that were helpful in the given situation than unsuccessful participants. A similar study using virtual city walks was performed by Nevelsteen (2013) using Google Street View instead of a real-world situation. The author in the study created two clusters of participants based on their navigation strategy, but the paper is focused more on the analysis procedure of the eye-tracking data than on actual results.

The previous studies were conducted to analyze the different map types that are used in presented studies. The most frequently analyzed maps contain perspective views of the terrain. The main goals of the chapter are:

1. To determine whether pseudo-3D visualization of buildings affects the ability to find a point symbol
2. To determine which of three visualizations of the extinct village will be the most effective for respondents

5.3 Method

5.3.1 Equipment

The eye-tracking technology is based on the principles of tracking human eye movements during perception of the visual scene. The measurement device is commonly known as an *eye-tracker*. Most modern eye-trackers measure the eye position and gaze direction using remote methods, relying on the measurement of the pupil and corneal reflection of a closely situated direct infra-red light source. The reflected light is recorded by a high-speed video camera. This information is then analyzed to extract eye movement and gaze direction from changes in corneal reflection.

The studies reported here used an eye-tracking device manufactured by SensoMotoric Instruments Company (model SMI RED 250). Within the study, data were recorded at a frequency of 120 Hz. Eye positions were recorded every 8 ms. The eye-tracker was supplemented by a web camera, which recorded participants' activity during the experiment. This video helped to reveal the possible cause of missing data, respondents' reactions to the stimuli, and their comments on the particular maps.

For data visualization and analyses, three different applications were used. The first was SMI BeGaze, which is software developed by the manufacturer of the device. The open-source software OGAMA and CommonGIS, developed at the Fraunhofer Institute in Germany, were also used for analysis of eye-tracking data.

5.3.2 Fixation Detection

The eyes move in a number of different ways, simultaneously responding to commands from a number of different brain areas. One of the most important types of eye movement is not really a movement at all, but rather, the ability to keep the eye trained on a relatively fixed spot in the world. This is known as *fixation*. Our visual experience is generally made up of a series of fixations on different objects. To get from one fixation to the next, the eyes make rapid, ballistic movements known as *saccades*. Saccadic movements are the fastest movements made by any part of the human body, with rotational velocities as high as 500 degrees/second for a large saccade. During saccades, perceivable events or objects are "invisible," not just because of the blur but due to a presumed neural process known as *saccadic suppression* (Hammoud & Mulligan, 2008).

During a fixation, eyes are relatively steadily looking at one small region in the visual field. Irwin (1992) states that the average fixation duration is between 150 and 600 ms. Based on statistical analysis of fixations, saccades, their mutual relationship, and other characteristics, it is possible to identify certain attributes of respondent behavior and find out what respondents are perceiving and what is their strategy for inspecting the stimuli.

A specific detection algorithm has to be used for eye-movement analysis. The development of the eye-movement classification algorithms has a long history. Almost every eye-movement classification algorithm has a set of input parameters that can significantly impact the result of classification. Numerous algorithms exist, but most are used for low–event rate data (up to 250 Hz). The algorithm used in the present studies is called *I-DT*. It takes into account the fact that successive fixations can be in close spatial proximity in the eye-movement trace (Salvucci & Goldberg, 2000). The algorithm defines a temporal window, which moves one point at a time. The spatial dispersion created by the points within this window is compared against a threshold. If dispersion is below the threshold, the points within the temporal window are classified as a part of fixation; otherwise, the window is moved by one sample, and the first sample of the previous window is classified as a saccade (Komogortsev & Khan, 2004).

For the studies reported in this chapter, the algorithm was instantiated using SMI BeGaze and OGAMA software. Threshold values in BeGaze were set to 80 ms for the "duration threshold" and 50 pixels for "dispersion threshold." In OGAMA, the most important parameters are "Maximum distance" and "Minimum number of samples," which correspond to dispersion and duration in BeGaze. Thresholds in OGAMA were set to 15 pixels (distance) and 10 samples. More information about these settings is presented in Popelka (2014a).

5.3.3 Experiment Design

In the first study dealing with non-photorealistic visualization, screenshots from three map portals were used as stimuli. The main problem was to control the carry-over (learning) effect, because the experiment was designed as within-subject. Carry-over effect refers to any lingering effects of a previous experimental condition that are affecting a current experimental condition. For that reason, fictitious point symbols were placed into the maps, so it was possible to change the position of points on each map in a pair. The same set of point symbols was placed in both maps in each pair, and the style of the points corresponded to the map.

In the second study, screenshots from a developed map application were used as stimuli. In some cases, similar views of the virtual village were used with a different type of visualization. To avoid learning effect, stimuli were presented in random order. The experiment also contained many images used mainly as distractors.

5.4 Study 1: Non-Photorealistic 3D Visualization

The aim of the study was to compare 2D and 3D visualizations of city maps during the search for specific points on the map. The aim of the experiment was to verify the following hypotheses:

- *H1*: In terms of the aesthetics, respondents would prefer the 3D map.
- *H2*: In terms of the suitability to the search task, respondents would prefer the 2D map.
- *H3*: Finding of the target will be difficult on stimulus with a tilted map.
- *H4*: Finding of the target will be easier on stimuli with a 2D variant of maps.

The first two hypotheses were defined according to the result of the previous study (Popelka, 2014b), in which similar answers were collected from the questionnaire dealing with tourist maps with and without shading. The other two hypotheses were defined on the basis of author's intuition, as it was not possible to adopt them from previous studies, because no such studies with similar material exist.

5.4.1 Participants

Forty participants (24 females, 16 males) volunteered to participate in the experiment. Twenty-seven of them were cartographers; 13 were not. The majority of participants were 20–25 years old. Before the experiment, they filled out the short demographic questionnaire and had to estimate how often they used Internet map portals such as Google Maps or OpenStreetMap. Most of the participants reported that they used these map portals every day.

5.4.2 Procedure

First, the eye-movement calibration was performed. Next, the participants were instructed that their task was to find one particular point symbol on the map as fast as possible and mark it with a mouse click. For each trial, a fixation cross was displayed for 500 ms before the stimulus was presented. Respondents had a maximum of 30 s to find the target. For most of the trials, the time was fully sufficient. At the end of the experiment, participants were asked to choose which version of the map was more suitable for finding the answer and which version they liked more.

5.4.3 Design

The study used a within-subject design with stimulus type as the independent variable. There were 18 stimuli (nine 2D and nine pseudo-3D) maps of cities. Although the pairs are numbered (1–9) in the chapter, stimuli were presented in random order.

5.4.3.1 Stimuli

The experiment contained nine static maps from three different Internet map portals (Google Maps, OpenStreetMaps [OSM], and F4map). The first one was the well-known Google Maps (Stimuli 1–5). For larger cities, the Google map contains a perspectival representation of blocks of buildings. The maps were simplified so as not to contain labels and original point symbols.

The second type of maps was maps from OpenStreetMap.org (Stimuli 6–8). In the default version, there existed no option to display a block of buildings in perspective view. However, due to the free availability of OSM data, there exist some possibilities for displaying pseudo-3D content. The well-known project osmbuildings.org, which is an additional layer to existing web maps, was used to create pseudo-3D maps.

The last stimulus used in the study (Map 9) also displays OSM data through the F4map portal. The F4map portal is available only in a beta version, but it automatically creates the 3D variant of cities across the world. In the previous examples (Google Maps and OSMbuildings), the pseudo-3D effect is created without deformation of the map. In the case of F4map, the map is tilted, and the more distant features are displayed as smaller than the closer ones. An example of each type of stimulus (Google Maps, OSMbuildings, and F4map) is shown in Figure 5.2.

One map in each pair is a standard 2D map. The other map in the pair contains a pseudo-3D visualization of buildings. The task was to find and mark (with a mouse click) a specific point symbol. These targets (point symbols) were imposed on each map. The same symbol set (i.e., fast food, university, theater, accommodation, etc.) and the same number of symbols were used for both maps in a pair to unify the search task. Each map contained only one correct answer (target point symbol). Across all of the stimuli, point symbols were placed at a similar distance from the center of the image (where respondents' gaze starts).

FIGURE 5.2
Pairs of 2D (left) and 3D (right) stimuli. (a) Google Maps (from http://maps.google.com), (b) OSMbuildings (from http://osmbuildings.org), (c) F4map (from http://map.f4-group.com).

5.4.4 Results

Results for nine pairs of stimuli from three different map portals were analyzed with the use of four eye-tracking measures and visual analysis using the FlowMap method. Moreover, the experiment was complemented by a short questionnaire focused on the subjective opinion of respondents about the analyzed maps.

Participants found the 2D map more suitable for answering the question (finding the point symbol on the map). The majority of participants preferred the 2D variant of the map. A relatively large number of participants chose the answer "It depends on the specific map." The distribution of the preference judgments was almost balanced, as shown in Figure 5.3.

For statistical analysis of eye-movement data, the following eye-tracking summary measures were calculated: Time to Answer (click), Fixation Count, Fixation Duration Median, and Scanpath Length. For all of these derived measures, median values were calculated. Medians were used because they are not affected by outliers. Data were analyzed with the use of the Wilcoxon rank sum test at the 0.05 significance level.

The first part of the statistical analysis was the comparison of the differences between 2D and 3D for all stimuli. The left-hand side of Table 5.1 shows that statistically significant differences were found for measures Time to Answer and Fixation Duration. Greater values of the median of both measures were observed for the 3D version of the maps. The biggest difference between 2D and 3D versions for all of the derived measures was found for the pair of stimuli No. 9 (the F4map). Without including the pair of stimuli No. 9 in the statistical comparison, no statistically significant difference for any of the monitored eye-tracking measures was found, as shown in the right-hand side of Table 5.1.

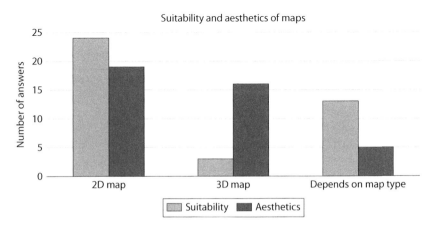

FIGURE 5.3
The results of the questionnaire focused on the subjective evaluation of the suitability and aesthetics of the maps of cities.

TABLE 5.1

Differences between 2D and 3D Variants of City Maps

	α	All Maps (Pairs 1–9)		All Maps except F4Map (Pairs 1–8)	
		W	*p*-Value	W	*p*-Value
Time to Answer	0.05	58,455.0	**.023**	48,345.0	.376
Fixation Count	0.05	59,532.5	.059	48,965.0	.339
Fixation Duration	0.05	58,952.5	**.036**	47,549.5	.118
Scanpath Length	0.05	62,603.0	.431	51,254.0	.982

Notes: Data were analyzed using the Wilcoxon test together for all stimuli. Statistically significant differences are in bold.
The W value for each group is calculated by subtracting the possible minimum rank that the group can take from the sum of the ranks, and the smallest W value is used for the test.

During a detailed analysis for each pair of maps separately, a statistically significant difference at the significance level $\alpha = 0.05$ was found for the majority of measures for two of the pairs of stimuli, numbered 7 and 9 in Table 5.2.

The largest number of statistically significant differences (7 of 9) was found for the Fixation Count measure. The biggest difference between 2D and 3D visualizations was found for all measures in the case of Stimulus No. 9. This result was expected, because the 3D Map No. 9 was tilted, which negatively affects the orientation in a map and made the task harder. The second largest value for Fixation Count was recorded for the 3D version of Map No. 5. This map, which displayed the downtown of New York with many 3D skyscrapers, was the most complex one from the set of "OSMbuildings." Visually, the difference between the 2D and 3D variants was quite high. For that reason, it is surprising that the difference between the 2D and 3D variants is so small in this case, as seen in Figure 5.4.

TABLE 5.2

Differences between 2D and 3D for All Pairs of City Maps

	Time to Answer	Fixation Count	Fixation Duration	Scanpath Length
Map 1	**0.009**	**0.009**	0.257	**<0.001**
Map 2	**0.048**	**0.023**	0.839	0.081
Map 3	**0.025**	**0.004**	0.750	**0.011**
Map 4	0.112	**0.042**	0.488	0.128
Map 5	0.365	0.603	**0.033**	0.656
Map 6	**0.007**	**0.008**	0.098	**0.003**
Map 7	0.051	**0.026**	**0.044**	**0.002**
Map 8	0.192	0.456	0.281	0.476
Map 9	**<0.001**	**0.002**	**0.025**	**0.018**

Notes: Data were analyzed using the Wilcoxon test for each stimulus separately. Statistically significant differences are in bold.

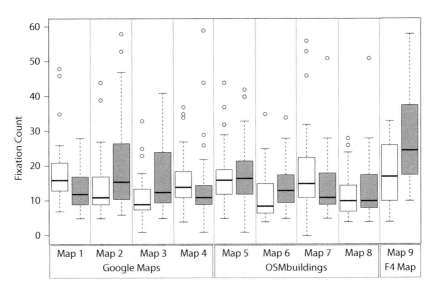

FIGURE 5.4

Fixation Count values for all nine pairs of stimuli in experiment "3D Cities." 3D version of the map is marked by gray color.

From the analysis of eye-tracking measures, it cannot be determined whether 2D or 3D versions of the maps were more suitable for solving the task. Larger values for the Fixation Count measure and the 2D version of the maps were observed for Maps 1, 4, and 7. For the other maps, the larger value was observed for the 3D version. This finding was obtained for all the other derived measures, with the sole exception of Fixation Duration. For this measure, statistically significant differences were observed, but only in three cases.

5.4.4.1 Visual Analytics

Eye-tracking data were visualized with the use of the FlowMap method in the V-Analytics software (Andrienko, Andrienko, Burch, & Weiskopf, 2012). The output of FlowMap shows aggregated gaze trajectories for all respondents between created Voronoi polygons, as shown by the purple arrows in Figure 5.5. In addition to the statistical evaluation, FlowMap shows the distribution of moves between fixations in the stimuli. Figure 5.5 shows FlowMap examples for pairs of stimuli No. 2, 5, and 7. All outputs have been created using the same settings. The width of arrows shows the total number of gaze movements between the Voronoi polygons. Only arrows representing more than three shifts are displayed.

For stimulus pair No. 2, the larger number of gaze movements was observed for the 3D version. In the case of stimulus pair No. 5, the number of aggregated

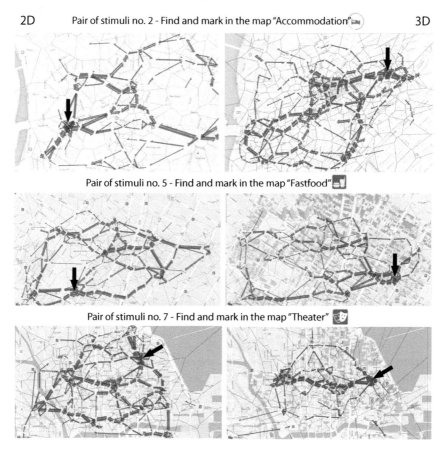

2D   Pair of stimuli no. 2 - Find and mark in the map "Accommodation"   3D

Pair of stimuli no. 5 - Find and mark in the map "Fastfood"

Pair of stimuli no. 7 - Find and mark in the map "Theater"

FIGURE 5.5
Visual analytics of selected tasks in CommonGIS. Black arrows point to the location of the target. 2D variant is on the left, 3D on the right.

trajectories over the 2D and 3D versions is comparable. In the case of stimulus pair No. 7, a larger number of trajectories was observed for the 2D version. These results are consistent with the statistical evaluation described earlier.

The hypothesis that respondents will prefer 3D maps in terms of aesthetics (H1) was not confirmed. Respondents did not have a clear aesthetic preference for the 2D or the 3D versions of the maps. However, the vast majority of participants chose the 2D map as the more suitable for the task (H2 verified). An analysis of four eye-tracking measures for each map pair showed that the biggest difference between 2D and 3D versions of maps was observed in the case of the pair of stimuli No. 9. Here, the 3D version was tilted, and it was difficult for the respondents to orient in the map (H3 verified). Values of eye-tracking measures were higher in this case, and respondents also reported the problem with orientation in the map during the testing.

In conclusion, we can say that the use of a pseudo-3D version of the map from the portal F4map (tilted) was not appropriate for the search task. For the other types of maps used (Google Maps and OSMbuildings), respondents sometimes preferred the 2D map and in other cases the 3D map. When all stimuli were analyzed together (except an F4map), it was not found that the type of visualization (2D, 3D) significantly influenced user perception. The hypothesis that finding the point symbol would be easier on a 2D variant of the stimulus (H4) was not verified.

5.5 Study 2: Photorealistic Visualization

For the evaluation of photorealistic 3D visualization, another eye-tracking experiment was created. The aim was to analyze how participants perceive different visualizations of an extinct village, Čistá, in the western part of the Czech Republic. For the study, data from a bachelor thesis (Dědková, 2012) were used.

5.5.1 Participants

Twenty-eight (19 male and nine female) individuals participated in the experiment. Eighteen respondents were students of geoinformatics, and the remaining 11 were students of other departments. The average age of participants was 23 years.

5.5.2 Design

The eye-tracking experiment consisted of three parts. Stimuli in the first part were views of the village presented in three versions: a cadastral map, an orthophoto, and an orthophoto overlaid with created 3D models of the village. These free-viewing stimuli (stimuli with no task) were displayed for 4 s, and respondents were not instructed to solve any particular task during their observation. In the second part of the experiment, respondents were tasked with finding a particular building in the village as fast as possible. All stimuli were presented in a random order, and it was a within-subject study. The independent variable was the type of image.

The aim of this study was to test these hypotheses:

- *H1*: The smallest number of fixations would be recorded for the 3D visualization maps.
- *H2*: The highest number of fixations would be recorded for the cadastral maps.
- *H3*: The best stimulus for the search for a specific building would be the one with a 3D map.

The first two hypotheses were defined according to the theory that respondents would perform more fixations on the stimuli with more objects (the cadastral map). The last hypothesis was inspired by the study of Lange (2001), who performed an empirical study in which respondents were asked to order a set of real images and computer-generated 3D maps. The results of the study showed that 75% of participants highlighted the very detailed 3D model with textures.

5.5.3 Stimuli

As stimuli in the second case study presented in this chapter, images from the web application created by Popelka and Dedkova (2014) were used. The web application shows a 3D model of the extinct village Čistá in the Czech Republic, which was destroyed during the creation of a military area in 1947.

Examples are shown in Figure 5.6. The cadastral map was the Imperial Imprint of Stable Cadastre from 1841 provided by the Czech Office for Surveying, Mapping and Cadastre (ČÚZK) in Prague. The orthophoto image of the Čistá village provided by the Military Geographical and Hydro-Meteorological Office in Dobruška (VGHMÚř) was acquired during 1947. The imagery was made before the destruction of the ruins of the village. That imagery was, of course, crucial for the creation of the 3D model of the village. It was created in Google SketchUp. More information about this model can be found in Popelka and Dedkova (2014).

5.5.4 Results

For the purpose of this chapter, results were analyzed for three stimuli, showing the three different visualizations of the southeast view of the village. Each image was shown to the respondents for 4 s. Two eye-tracking measures were evaluated—Fixation Count and Fixation Duration.

As seen from Figure 5.7, the greatest number of fixations with the shortest length was observed in the case of the cadastral map. A greater number of

FIGURE 5.6
Stimuli showing a view of Čistá village from the southeast in three variants.

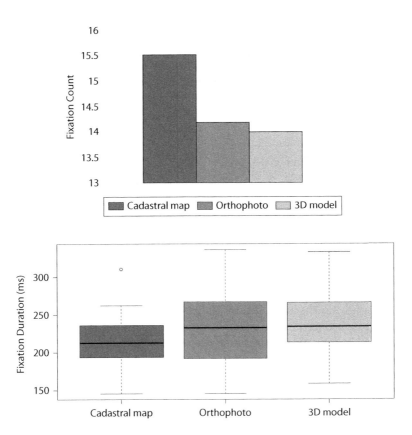

FIGURE 5.7
Plots showing the average Fixation Count and Fixation Duration for 28 respondents.

shorter fixations indicates a less efficient search (Jacob & Karn, 2003). For the other two versions of the maps, the values for the measures were similar. The number of fixations was significantly lower (using Wilcoxon rank sum test) than in the cadastral map. This fact indicates a more efficient search (Jacob & Karn, 2003).

Data were statistically analyzed by analysis of variance (ANOVA) and the Tukey honest significant difference (HSD) test. Because the residuals of the ANOVA did not have a normal distribution, the correctness of results was confirmed with bootstrapping. Bootstrapping was done as a modification of Manly's approach—Unrestricted Permutation of Observations (Manly, 2006). Statistically significant differences (again at the significance level $\alpha = 0.05$) were observed for the Fixation Duration measure between the cadastral map and the 3D model, as shown in Table 5.3. These results show that respondents' eyes roamed from place to place in the case of the cadastral map and that the 3D model was the most useful tool for the visual search.

TABLE 5.3

Results of ANOVA for Fixation Duration for the Three Types of Visualization
(Cadastral Map, Orthophoto, and 3D Model)

Fixation Duration	Diff	Lwr	Upr	p adj
Orthophoto—Cadaster	19.499	−6.802	45.800	.186
Cadaster—3D model	28.178	1.877	54.479	**.033**
3D Model—Orthophoto	8.679	−17.622	34.980	.712

Notes: Diff means Difference in the observed means; Lwr means the lower end point of the
interval; Upr means the upper end point; p adj means the p-value after adjustment for the
multiple comparisons.

5.5.4.1 Visual Analytics

The task in the second part of this experiment was to find a particular building in the extinct village of Čistá. The pair of stimuli (orthophoto image and
3D model) were presented.

The task, in this case, was to find a school building as quickly as possible.
For data analysis, the areas of interest (AOI) method was used. Areas of interest were marked around all major locations on both stimuli, as illustrated in
Figure 5.8. Red circles with numbers represent the number of recorded fixations of all 28 respondents in individual AOIs. The width of yellow arrows
represents the number of transitions between AOIs.

On the orthophoto in Figure 5.8a, the greatest number of fixations was
obtained in the AOI covering a block of buildings in the southern part of the
village. The relatively large number of fixations (23) was recorded around
the pond at the bottom of the stimulus. For the 3D model stimulus, only four
fixations were measured around the pond. In this case, most of the fixations
were recorded near City Hall (52) and the school (41). From this distribution

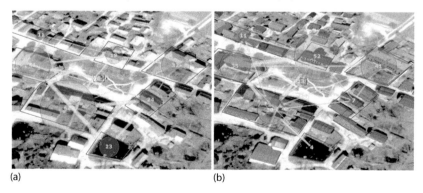

(a)   (b)

FIGURE 5.8

Results of the AOI analysis of orthophoto (a) and 3D model (b) stimuli. The task was to find the
"School" as fast as possible. Red circles with numbers represent the total number of fixations
recorded in the particular AOI. The width of yellow arrows represents the number of fixation
transitions.

of fixations, it is evident that respondents considered these two buildings as important. From the stimuli, it was not possible to identify which one was the school. However, the fact that respondents focused on these two buildings shows that the 3D visualization was better than the display aerial image for solving the task. The total number of fixations for the orthophoto stimulus was lower (285) than for the 3D model stimulus (371).

In conclusion, it is possible to say that in the case of free-viewing the stimuli, the smallest number of the longest fixations was found in stimuli with the 3D model (H1 validated). In contrast, the cadastral map stimulus contained a higher number of fixations, because respondents' eyes were roaming from place to place or could be affected by text labels on the map. A statistically significant difference was found between these two stimuli for Fixation Duration (H2 verified). An analysis of AOI for the second task indicated that the 3D model was more suitable for the task of finding the particular building in the village (H3 verified).

5.6 Discussion and Conclusion

User studies focused on 2D and 3D visualizations of urban areas were analyzed with the use of the eye-tracking device. The advantage of analyzing eye movements instead of the widely used questionnaire method is that from eye-movement data, it is possible to analyze user reactions and their strategy during solving the task, not only their answer.

Two case studies of the perception of 2D and 3D non-photorealistic and photorealistic visualization of urban areas were performed. In the first study, a respondent's task was to find a particular point symbol as fast as possible and mark it with a mouse click. The purpose of the study was to find out whether the 3D visualization would negatively influence the search for a specific point symbol. It was problematic to distinguish any explicit deliberative strategies. No significant differences between 2D and 3D maps were found for four measures (Time to Answer, Fixation Count, Fixation Duration Median, and Scanpath Length). In some cases, answers were faster in the 2D map and in some cases, in the 3D one.

The only exception was F4map, where statistically significant differences were observed for all recorded eye-tracking measures. This type of map should not be used very often, because users have problems with orientation in it. The results for all other stimuli (Google Maps and OSMbuildings) indicate that in situations when it is reasonable and desirable, the 3D map of the city could be used instead of the standard 2D map. In the 3D map, more information is contained, and the 3D representation did not influence the reading of the map and its comprehensibility.

Data from the short questionnaire presented after the experiment shows that respondents consider the 2D variant more suitable for answering the question. Respondents did not clearly favor one of the map design variants from the aesthetic point of view.

In the free-viewing part of the second case study, the greatest number of fixations was observed for the cadastral map and the smallest for the 3D model. Respondents' eyes roamed from place to place in the case of the cadastral map, and the 3D model was the most useful tool for the visual search. In the next part of the experiment, a pair of stimuli (orthophoto vs. 3D model) was presented, and respondents had to find "school" as fast as they could. From AOI analysis, it was clear that the 3D model was more useful for this task than the orthophoto.

The search for a specific building was faster for a 3D map. These results could indicate that 3D visualization is better than 2D, but this cannot be generalized to any 3D visualization and any task. If the map or the task were different, it is possible that the results would differ too. For that reason, more testing is necessary.

The present studies were focused on the single aspect of 3D visualization. More testing with different 3D visualizations and different tasks will be necessary to develop generalized findings. More complex tasks requiring the imagination of 3D objects in the map will help to better understand the possibilities and limitations of 3D visualization.

It is problematic to compare the results with other studies, because 3D urban maps have never been investigated previously. Savage et al. (2004) analyzed 2D and 3D visualization of contour maps, and they did not find any advantage of 3D maps for those tasks requiring elevation information. The results of the questionnaire study of Petrovič and Mašera (2004) showed that the draped topographic (3D) map has been recognized as being almost as adequate for height or direction measurement as the traditional 2D map.

The results of both case studies suggest that for visualization of urban areas, 3D visualization can be used. Either it did not negatively influence map reading (as was seen in the first case study) or it was even better than the 2D variant (as was seen in the second case study). As Wood et al. (2005) stated, improvements in our understanding of the process of cognition of 3D scenes may help us to construct more effective 3D realizations. The author believes that the presented case studies will help in this understanding.

Acknowledgment

This research was supported by the project of Operational Program Education for Competitiveness—European Social Fund (project CZ.1.07/2.3.00/20.0170 of the Ministry of Education, Youth and Sports of the Czech Republic).

References

Andrienko, G., Andrienko, N., Burch, M., & Weiskopf, D. (2012). Visual analytics methodology for eye movement studies. *IEEE Transactions on Visualization and Computer Graphics*, 18(12), 2889–2898.

Bleisch, S. (2012). 3D geovisualization—definition and structures for the assessment of usefulness. *ISPRS Annals of the Photogrammetry, Remote Sensing and Spatial Information Sciences*, I–2, 129–134.

Cartwright, W. E. (2006). Using 3D models for visualizing "The city as it might be". *Proceedings of ISPRS Technical Commission II Symposium*. Vienna, Austria.

Dědková, P. (2012). *3D vizualizace zaniklé obce a její hodnocení z hlediska uživatelské kognice. (Bc.)*. Olomouc: Univerzita Palackého v Olomouci.

Döllner, J. (2007). *Non-Photorealistic 3D Geovisualization Multimedia Cartography* (pp. 229–240). Berlin/Heidelberg, Germany: Springer-Verlag.

Döllner, J. (2008). *Visualization, Photorealistic and Non-photorealistic Encyclopedia of GIS* (pp. 1223–1228). Berlin/Heidelberg, Germany: Springer-Verlag.

Döllner, J., & Buchholz, H. (2005). Non-photorealism in 3D geovirtual environments. *Proceedings of AutoCarto*. Las Vegas, NV.

Durand, F. (2002). An invitation to discuss computer depiction. *Proceedings of the 2nd International Symposium on Non-Photorealistic Animation and Rendering*. Annecy, France.

Fuhrmann, S., Komogortsev, O., & Tamir, D. (2009). Investigating hologram-based route planning. *Transactions in GIS*, 13(s1), 177–196.

Greenberg, S., & Buxton, B. (2008). Usability evaluation considered harmful (some of the time). *Proceedings of the SIGCHI Conference on Human Factors in Computing Systems*. Florence, Italy.

Guidi, G., Frischer, B., & Lucenti, I. (2007). Rome reborn—virtualizing the ancient imperial Rome. Paper presented at the Workshop on 3D Virtual Reconstruction and Visualization of Complex Architectures.

Haber, R. B., & McNabb, D. A. (1990). Visualization idioms: A conceptual model for scientific visualization systems. *Visualization in Scientific Computing*, 74, 93.

Haeberling, C. (2002). 3D map presentation—A systematic evaluation of important graphic aspects. Paper presented at the ICA Mountain Cartography Workshop, Timberline Lodge, Mt. Hood, OR.

Hammoud, R. I., & Mulligan, J. B. (2008). Introduction to eye monitoring. In R. I. Hammoud (Ed.), *Passive Eye Monitoring: Algorithms, Applications and Experiments* (pp. 1–19). Berlin/Heidelberg, Germany: Springer.

Irwin, D. E. (1992). Visual memory within and across fixations. In K. Rayner (Ed.), *Eye Movements and Visual Cognition: Scene Perception and Reading* (pp. 146–165). New York: Springer.

Jacob, R. J., & Karn, K. S. (2003). Eye tracking in human-computer interaction and usability research: Ready to deliver the promises. *Mind*, 2(3), 4.

Jedlicka, K., Cerba, O., & Hajek, P. (2013). Creation of information-rich 3D model in geographic information system—case study at the Castle Kozel. *13th International Multidisciplinary Scientific Geoconference, SGEM 201.3*. Albena, Bulgaria.

John, M. S., Cowen, M. B., Smallman, H. S., & Oonk, H. M. (2001). The use of 2D and 3D displays for shape-understanding versus relative-position tasks. *Human Factors: The Journal of the Human Factors and Ergonomics Society*, 43(1), 79–98.

Kiefer, P., Giannopoulos, I., & Raubal, M. (2014). Where am I? Investigating map matching during self-localization with mobile eye tracking in an urban environment. *Transactions in GIS*, 18(5), 660–686.

Kiefer, P., Straub, F., & Raubal, M. (2012). Location-aware mobile eye tracking for the explanation of wayfinding behavior. *Proceedings of the AGILE'2012 International Conference on Geographic Information Science*. Avignon, France.

Komogortsev, O., & Khan, J. (2004). Predictive perceptual compression for real time video communication. *Proceedings of the 12th Annual ACM International Conference on Multimedia*. New York, NY.

Konečný, M., Kubíček, P., Stachoň, Z., & Šašinka, Č. (2011). The usability of selected base maps for crises management—users' perspectives. *Applied Geomatics*, 3(4), 189–198.

Koutsoudis, A., Arnaoutoglou, F., & Chamzas, C. (2007). On 3D reconstruction of the old city of Xanthi. A minimum budget approach to virtual touring based on photogrammetry. *Journal of Cultural Heritage*, 8(1), 26–31.

Kraak, M. (1988). *Computer-assisted Cartographical 3D Imaging Techniques*. Delft, The Netherlands: Delft University Press.

Lam, H., Bertini, E., Isenberg, P., Plaisant, C., & Carpendale, S. (2012). Empirical studies in information visualization: Seven scenarios. *IEEE Transactions on Visualization and Computer Graphics*, 18(9), 1520–1536.

Lange, E. (2001). The limits of realism: Perceptions of virtual landscapes. *Landscape and Urban Planning*, 54(1), 163–182.

Manly, B. F. (2006). *Randomization, Bootstrap and Monte Carlo Methods in Biology* (Vol. 70). Boca Raton, FL: CRC Press.

Montello, D. R. (2009). Cognitive research in GIScience: Recent achievements and future prospects. *Geography Compass*, 3(5), 1824–1840.

Nevelsteen, K. (2013). Attention allocation of traffic environments of international visitors during virtual city walks. Paper presented at the ET4S 2013, Leuven, Belgium.

Petrovič, D., and Mašera, P. (2004). Analysis of user's response on 3D cartographic presentations. Paper presented at the 7th Meeting of the ICA Commission on Mountain Cartography. Bohinj, Slovenia.

Popelka, S. (2014a). Optimal eye fixation detection settings for cartographic purposes. Paper presented at the 14th SGEM GeoConference on Informatics, Geoinformatics and Remote Sensing. Albena, Bulgaria.

Popelka, S. (2014b). The role of hill-shading in tourist maps. *CEUR Workshop Proceedings*. Vienna, Austria.

Popelka, S., & Brychtova, A. (2013). Eye-tracking study on different perception of 2D and 3D terrain visualisation. *Cartographic Journal*, 50(3), 240–246.

Popelka, S., & Dedkova, P. (2014). Extinct village 3D visualization and its evaluation with eye-movement recording. *Lecture Notes in Computer Science (Including Subseries Lecture Notes in Artificial Intelligence and Lecture Notes in Bioinformatics)*, Vol. 8579 LNCS (pp. 786–795).

Popelka, S., & Dolezalova, J. (2016). Differences between 2D map and virtual globe containing point symbols an eye-tracking study. *Informatics, Geoinformatics and Remote Sensing Conference Proceedings, Sgem 2016*, III, (pp. 175–183). Albena, Bulgaria.

Putto, K., Kettunen, P., Torniainen, J., Krause, C. M., & Tiina Sarjakoski, L. (2014). Effects of cartographic elevation visualizations and map-reading tasks on eye movements. *The Cartographic Journal*, 51(3), 225–236.

Remondino, F., El-Hakim, S., Girardi, S., Rizzi, A., Benedetti, S., & Gonzo, L. (2009). 3D virtual reconstruction and visualization of complex architectures— The "3D-ARCH" project. Paper presented at the ISPRS Working Group V/4 Workshop 3D-ARCH. Trento, Italy.

Remondino, F., & Rizzi, A. (2010). Reality-based 3D documentation of natural and cultural heritage sites—techniques, problems, and examples. *Applied Geomatics*, 2(3), 85–100.

Salvucci, D. D., & Goldberg, J. H. (2000). Identifying fixations and saccades in eye-tracking protocols. *Proceedings of the 2000 Symposium on Eye Tracking Research and Applications*. Palm Beach Gardens, FL.

Savage, D. M., Wiebe, E. N., & Devine, H. A. (2004). Performance of 2D versus 3D topographic representations for different task types. Paper presented at the Human Factors and Ergonomics Society Annual Meeting.

Schobesberger, D., & Patterson, T. (2007). Evaluating the effectiveness of 2D vs. 3D trailhead maps. *6th ICA Mountain Cartography Workshop Mountain Mapping and Visualization*, pp. 201–205. Lenk, Switzerland.

Simone, F. D., Presta, R., & Protti, F. (2014). Evaluating data storytelling strategies: A case study on urban changes. Paper presented at the COGNITIVE 2014, Venice, Italy.

Slocum, T. A., Blok, C., Jiang, B., Koussoulakou, A., Montello, D. R., Fuhrmann, S., & Hedley, N. R. (2001). Cognitive and usability issues in geovisualization. *Cartography and Geographic Information Science*, 28(1), 61–75.

Upson, C., Faulhaber Jr, T. A., Kamins, D., Laidlaw, D., Schlegel, D., Vroom, J., & Van Dam, A. (1989). The application visualization system: A computational environment for scientific visualization. *IEEE Computer Graphics and Applications*, 9(4), 30–42.

Wilkening, J., & Fabrikant, S. I. (2013). How users interact with a 3D geo-browser under time pressure. *Cartography and Geographic Information Science*, 40(1), 40–52.

Wood, J., Kirschenbauer, S., Döllner, J., Lopes, A., & Bodum, L. (2005). Using 3D in visualization. In J. Dykes (Ed.), *Exploring Geovisualization* (pp. 295–312). Oxford: Elsevier.

6

Designing Geographic Information for Mountains: Mixed Methods Research

Raffaella Balzarini and Nadine Mandran

CONTENTS

KEYWORDS: *map use, mental representation, qualitative and quantitative data analysis, mobile eye-tracking techniques, map design, mountain cartography*

6.1 Introduction

This chapter presents a tutorial for an exploratory experimental approach based on qualitative and quantitative methods of data collection and data analysis, which considers the qualitative versus quantitative distinction as a complementarity rather than an opposition. The starting question for any experimental approach is often: When is it appropriate to use qualitative techniques, and when are quantitative techniques more suitable to my study? As Libarkin and Kurdziel (2002) explain, "an effective qualitative research allows outsiders to view a situation from the perspective of the individuals involved, while quantitative research, on the other hand, typically focuses on overarching 'truths' that are applicable to a range of similar settings and populations" (p. 80).

The research discussed in this chapter illustrates the success that comes from combining statistical analysis with contextual data. We conducted a geo-cognitive study of geographic information (GI) in mountain cartography in the particular case of ski resorts mapping. Cognitive research in geographic information science (GIScience) includes the study of both internal mental and external symbolic structures and processes and is practically motivated by the desire to improve the usability, efficiency, equity, and profitability of geographic information and geographic information systems (GIS) (Montello, 2009). According to this approach, our study fits with the design of information displays (geovisualization), which is the clearest case of cognitive research within GIScience. This continues the tradition of studying cognitive aspects of cartographic communication (Lloyd, 1993; MacEachren, 1995; Montello, 2002). Our study and approach also fit with human factors research, which has looked at the nature of reasoning and decision making with geographic/cartographic information (Çöltekin, Fabrikant, & Lacayo, 2010; Montello, 2009).

Cartographic user research relies on simultaneous or sequential mixing methods. Mixing sequentially refers to the situation when different experiments are conducted and their outcomes are linked to each other. Typically, the separate experiments focus on different aspects of this main objective, each using the most appropriate method to study the specific research question at hand. Simultaneous mixing refers to situations when data from multiple methods are gathered at the same time: for example, explaining while the participant's eye movements are being recorded (Ooms, 2016).

The chapter is organized as follows. The Context section introduces the case study's context, its background, and the overall resort design issue. The Methods section describes the approach applied to the study and details the experimental protocols and the results. The Discussion section considers conclusions as well as the outlook for future studies.

6.2 Context

The geo-cognitive study we present in this chapter is part of a multidisciplinary research project titled MEmory, COgnition and MOdeling of mountain landscape (MECOMO). The project is being carried out by historians, computer scientists, geographers and cognitive scientists of the Grenoble Alps University and is focused on the representation of ski resort trail maps. The general goal of the study is to provide insights about the effectiveness of panoramic maps and more broadly, to suggest innovative and efficient representation of the geographic information in mountainous terrain, to respond more adequately to the new practice in winter sports. The MECOMO project is coordinated by the International Society of Alpine History and the Commission on Mountain Cartography of the International Cartography Association.

In the semiotic categorization of mountain panoramas (Patterson, 2000; Tait, 2010), trail maps for ski areas are iconic images used in the sport of skiing. They express the relation of skiing to the terrain on which ski resorts are built. The combination of larger areas to map and increases in sport finance led to a shift from simple wayfinding maps to more elaborate mountain portrayals (Patterson, 2000). Artists have employed three main types of views for ski mountain resorts: planimetric views, profile views, and panoramic views. Planimetric views are images viewed from directly above the skiing area. Profile views are generally very simple elevation views of the mountain that have little or no three-dimensional (3D) character or photogrammetric perspective. Panoramic views are by far the most common type of map to depict skiing trails. They are oblique perspectives from some given angle and are often not topographically accurate (Tait, 2010).

Panoramas respond to the desire for resorts to look impressive to potential visitors. Ski areas often ask for the mountain to "look bigger" and for the artist to distort the mountain for this purpose. This request can pose a serious challenge, which in some cases is resolved by local distortion of the mountain terrain and camera view (Patterson, 2000). Landscape artists have addressed such challenges by applying local deformation to map objects in hand-painted hiking and skiing panoramas. Using digital means, the painters' techniques may be translated into geometry deformation algorithms for digital panorama creation. Based on Berann's landscape manipulation techniques (Patterson, 2000), this solution should help the cartographer to deform a digital terrain model by intuitively manipulating the surface in a 3D display (Jenny, Jenny, Cartwright, & Hurni, 2011). The digital imitation of the techniques of panorama painters has also been improved by combining local terrain deformation and progressive bending algorithms with painterly rendering methods for panoramas (Bratkova, Shirley, & Thompson, 2009).

We investigated the evolution of resort maps and of the use of ski maps by skiers, both according to a cognitive perspective. Previous evaluations of the effectiveness of panoramic maps (Spengler & Räber, 2012) have not considered the cognitive-perceptual aspects of map understanding and also were concerned with maps of urban areas rather than the natural environment. That said, we did adopt the method of studying map understanding via eye-tracking (as in Kiefer, Giannopoulos, & Raubal, 2014).

Our study concerns mountain representations according to the artistic style of the "Atelier Pierre Novat," the pioneer French ski map-maker (Belluard, Nova, & Novat, 2013), who developed the panoramic view. Artists who follow the Atelier Pierre Novat tradition apply their skill to combine in a single mountain panorama objects of the landscape that are not visible in any single real angle of view. The panorama is realistic and easy to read but is subtly distorted. Figure 6.1 shows an example ski trail map of the Alpes d'Huez ski resort in the French Alps; it was made by Arthur Novat using the Atelier Pierre Novat style.

6.3 Research Issue

Since the 1930s, ski resorts and tourist and sporting innovations for mountain economies have needed cartographic representations in print, bounded by fixed dimensions. According to some mountain operators,

FIGURE 6.1
Alpes d'Huez ski trails maps. (Courtesy of Atelier Pierre Novat.)

the information provided by paper ski maps no longer meets the needs of many customers; the question now arises of their adaptation to new digital mapping practices (e.g., iPhone, smart tablets). At present, ski paper maps coexist with digital high technology (GIS, mobile apps, and digital 3D ski maps). Ski maps are an assembly of numerous graphic-geographic symbolic objects: ski trails and colored slope tracings, vegetation, peaks, rocks, shadows and light, buildings, and so on. For all the forms of ski maps, there is an issue in the representation of the geographic information: What has to be represented and how? This can be posed as our two main research questions: (1) What geographic information (and its representation) makes ski paper maps effective to perform a user-skier's task? (2) What geographic information (and its representation) generates misunderstanding and difficulties in a user-skier's comprehension of ski maps?

These questions determined our choice of research methods.

6.4 Our Research Approach

Our qualitative analysis relies on theoretical concepts of mental representation. Human activity is based on personal reconstructions of reality based on perceptual systems and previous knowledge (Gentner & Stevens, 1983). The internal representation requires spatial skills such as visualization, orientation, and spatial relations (Hegarty, Montello, Richardson, Ishikawa, & Lovelace, 2006). The external spatial representation refers to the organization, interpretation, and communication of information with maps, charts, or images. In ski map interpretation, mental representations refer to the natural and anthropogenic elements (objects) depicted in the ski map. Our study investigated the user's capacity to recognize these objects and to extract the relevant information. A preliminary step to study user activity was to get a deep understanding of how the expert-artist creates objects and information on a ski map. For this, the expert-artist's mental representations, and the resulting graphic objects, are known from previous works (Balzarini, Dalmasso, & Murat, 2015).

In our Experiment 1, mental representations were identified through the collection of data from a "thinking aloud" task followed by protocol analysis (Chi, 1997; Ericsson, 2006; Newell & Simon, 1972). In the think-aloud task, the participants are asked to verbalize the information they attend to while solving a problem. The purpose is to capture the processes of solving a problem or making a decision (i.e., conducting some task). In the data analysis, the participants are asked to verbalize explanations, descriptions, justifications, and rationalizations. The focus is to capture the mental model, or the knowledge that a solver/user has. In our case study, knowledge of the practical application of spatial analysis concepts and cartographic design was represented. The complementarity of both techniques provides data corresponding to all the concepts, objectives, roles, and relationships mobilized by a participant when performing a task. These techniques are commonly used in psychology to reconstitute a cognitive structure; they apply both to the study of expertise and to the study of user behavior (see Crandall, Klein, & Hoffman, 2006).

Once the skier's knowledge about ski maps had been elicited, our second experimental step consisted in studying perception and reasoning as a matter of *visual attention* using eye-tracking techniques (Ware, 2008). This allows the identification of ski map information that is most used (gazed at) during the ordinary task of viewing and interpreting a map in a skiing day (our Experiment 2). The visual exploration of a map involves the exploration of information in a visual scene, reasoning about the task at hand, and performing a visual search (Kiefer, Giannopoulos, Raubal, & Duchowski, 2017). Eye-tracking allows us to measure an individual's visual attention, yielding information on where, when, for how long, and in which sequence certain information in space or about space is looked at. Eye-tracking has proved to be a useful method for investigating how people interact with geographic information in decision situations (Kiefer et al., 2017).

Based on the results of the first two experiments, the third was a quantitative, large-scale survey whose main purpose was to provide statistical feedback on the use of ski maps and on their functions.

6.5 Experiment 1

6.5.1 Map Interpretation and Information Inventory

6.5.1.1 Method

Skiers' knowledge and reasoning can be expressed as a set of propositions, a set of concepts, a set of goals, or a set of rules from "what" the participant said (the content). However, to explore the elicited knowledge, the researcher must then assess the relations among the concepts and rules. Our research participants were asked to verbally identify the information to which they attended while solving a typical ski day's situation problem. Example problems are finding the fastest way to get to the ski school or finding a panoramic point. Participants then expressed their explanation and justification for their information search. Our research goal was to capture the

Experimental Questions: What are the skier's processes of solving a problem or making a decision in the ski resort? What knowledge does the skier have? What concepts and map features/representations does the participant verbalize while reading the ski map? What depictions are hard to interpret?

6.5.1.1.1 *Hypotheses*

> *H1*: Participants verbalize the information corresponding to the objects described in the legend but also some more elaborate information, some concepts (e.g., the natural limits of the ski areas are not traced on the ski map or indicated in the legend but can be recognized).
>
> *H2*: Maps contain information that can be difficult to interpret (e.g., combes, high rock bars, slope gradients, etc.)

6.5.1.1.2 *Participants*

The 20 participants were between 18 and 65 years old, 12 women and eight men. They were assigned into three groups according to their skiing experience level: two in the novice skier group (N), 10 in the intermediate group (I), and eight in the advanced group (A). Ski proficiency level was determined by self-evaluation according to the French Ski School standard.

6.5.1.1.3 *Independent Variable*

The participants were divided into two groups according to which of two different resort ski maps they would see: Alpes d'Huez (see Figure 6.1) or Villard de Lans, both situated in the French Northern Alps. These two ski resorts were chosen because they are representative of the regional ski resorts in terms of size (large and medium) and attendance. The participants were not familiar with these ski resorts.

6.5.1.1.4 *Tasks*

The participants were asked to perform two tasks. In Task 1 (T1), the participants were instructed to prepare their skiing day by exploring the ski resort's features on the ski map, explaining which areas of the ski resort they would like to discover and why. The participants were instructed to verbalize salient elements in the map that caught their attention (map objects, focal points) and then circumscribe them on the paper map using a felt-tip pen. Task 2 (T2) was a decision-making task. The participants were instructed to create a route between two fixed locations, trace it on the map with a felt-tip pen, and explain their reasoning about the path. Figure 6.2 shows a participant during the experiment.

6.5.1.2 **Results**

Using the methods of protocol analysis to segment the participants' utterances (e.g., Chi, 1977), we searched for tag phrases and keywords to form meaningful groups. Categories included:

- *Geomorphology*: References to features of the landscape and its morphology (e.g., hollow, cliff, tree, peak)
- *Geography*: Expressions relating to localization, orientation, or positioning

FIGURE 6.2
Participant during the verbal data experience.

- *Tracings*: References to the drawn lines (e.g., ski lift lines, ski run geometry)
- *Nomenclature*: References to names and numbers (names of ski runs and lifts, elevation values, place names)
- *Structures*: Roads, buildings, villages, etc.
- *Critique*: Expressions of incomprehension, uncertainty, or interpretation difficulties (e.g., need to zoom in; illegibility of map elements)

For example, the expression "There I *do not know what it looks like*, it seems to be [a] *cliff*" includes an expression of uncertainty and a geomorphologic descriptor. The expression "I'm looking for the info on the *ski lifts*. First, I try to *locate myself*" includes an expression of a tracing and a geographic reference. The expression "I look at the *colors of the ski run* (t2), but … here what interests me considering that it is *very beautiful is to enjoy a maximum with all varieties of landscape*" includes a reference to ski lift lines and a reference to the enjoyment of the landscape. Ski maps can facilitate (or not facilitate) the skier's achievement of their objectives. The results from the protocol analysis enabled us to describe the main goals skiers have in map comprehension, the difficulties they encounter in ski map comprehension, and the moments when the map provides a real support for decision making. The protocols included many expressions of the skiers' goals—explanations of why information is used (e.g., to find the connections of the trails network; to assess the quality of the snow). Map interpretation goals included:

- Evaluate the extension and organization of the ski area
- Define a departure point
- Define a point of arrival
- Assess the landscape and pleasant places
- Assess the difficulty of the ski runs and the length and capacity of ski lifts
- Find connections in the trails network
- Find break places
- Find panorama places
- Find easy ways to get to the starting point

Skier decision making and route planning is a process that is not without uncertainties and difficulties. Difficulties are manifested by misunderstanding, inconsistencies, missing details, and illegibility. We listed them by category and by the graphic objects concerned. Examples are presented in Table 6.1. The reference to color-coded ski runs has to do with the fact that standard colors are used to indicate run difficulty.

The difficulties were quantified according to the frequency of occurrence in the protocols of the participants. From a total of 55 expressions of difficulties identified in the utterances, 31% were related to geomorphology, 27% to

TABLE 6.1

Example Protocol Statements Illustrating Sources of Difficulty in Map Interpretation

Statement	Reference	Participant Experience Level
"Up [there] it is complicated to identify. [It] is very flat and the plan does not allow [me] to understand well"	Peaks and ridges	Advanced
"And there the vertical drop allows me to go there? I'm not quite sure. I have a doubt. Here I do not see well enough"	Slopes	Intermediate
"It was a shadow where we imagine a hollow so it's a little weird"	Hollows	Intermediate
"It would be good to have the contour and IGN map you would see right off the rocky ridge"	Rocks, cliffs	Advanced
"This is a blue run. But then I do not know if it goes down or it rises"	Colored ski runs	Advanced
"What seems strange is that if it's a green it cannot be so steep"	Colored ski runs	Intermediate
"It's always the difficult part on a ski map, to understand whether in fact it connects, if it goes up or goes down …"	Ski lift	Advanced

tracing, 18% to structure category, 16% to geography, and 7% to nomenclature. Deeper analyses of difficulties related to the geomorphic and tracing categories showed that all the participants expressed uncertainty ["I'm not sure that ..."], 44% of the participants expressed a legibility problem ["Not easy to see or to read ..."], 33% of the participants expressed a problem in understanding ["I don't know ..."], and 22% of participants referenced a perceived inconsistency in the ski map ["It's a little weird ..."].

Both of our hypotheses for Experiment 1 were verified. Participants' think-aloud verbalizations included references to explicit map information but also referenced their knowledge and goals (H1) (e.g., the natural geobiological boundaries of the ski areas are not traced on the ski map, nor are they indicated in legend, but they can be recognized by experienced skiers). In addition, maps have numerous features and graphic elements that are ambiguous or otherwise difficult to interpret (H2) (e.g., the actual gradient of slopes).

These results comprise an inventory on skiers' map interpretation concepts and goals and highlight some difficulties in comprehending and interpreting ski maps with respect to goals. To get deeper insights into the map interpretation process, our Experiment 2 involved the capture of data on eye movements.

6.6 Experiment 2

6.6.1 Eye-Movement Study

Eye-trackers measure a person's gaze, which is often interpreted as an indicator of visual attention. It is generally assumed that perception or comprehension take place only if gaze remains local to a region or a graphic object for a minimum amount of time, scaled in milliseconds. Thus, gazes are often aggregated spatio-temporally as *fixations*. A transition between two fixations is called a *saccade*, which is a rapid movement of the eye. The fixation period is generally between 200 and 600 ms, and saccades typically take between 20 and 100 ms. A comprehensive overview on eye-tracking technology and methodology can be found in Duchowski (2007), Holmqvist et al. (2011), and Kiefer et al. (2014). Eye-tracking studies for cartographic stimuli have been conducted since the 1970s. Recent work has focused on usability aspects of interactive maps, such as effectiveness and efficiency of different map designs (Çöltekin et al., 2010). The design of a map and the usability of the system in which it is shown influence wayfinding and self-localization (Kiefer et al., 2014).

Experimental questions: What areas of a ski map are fixated by skiers? What are the graphic objects most often gazed on, and why? In what ways are aspects of map interpretation difficulty manifested in gazes and fixations?

6.6.1.1 Hypotheses

H1: The area of a ski map that will be gazed at the most will be the central region of the map. The most explored area of the ski map is the central one.

H2: The graphic objects gazed at the most will be those in the tracing and the geomorphic categories, since these are more directly pertinent to decisions concerning the preferred ski route.

H3: The areas showing a misleading or ambiguous representation of the terrain will pose difficulties in interpretation, which will be manifested in gaze patterns.

6.6.1.2 Participants

The participants were five men and five women between 25 and 55 years old. They were assigned into three groups according to their ski level. Two were rated as Novice (N), five were rated as Intermediate (I), and three were rated as Advanced (A). Ski proficiency level was determined by self-evaluation according to the French Ski School standard.

6.6.1.3 Task

The participants were observed in a controlled laboratory and were asked to perform two tasks that are representative of ordinary skiing days. First, to prepare their skiing day, skiers had to explore the ski resort features depicted on the ski map to determine which areas they would like to discover and why (Task 1). Then, they created a route between two fixed locations and explained their reasoning and rationale (T2). Each participant's session took about 15 min.

6.6.1.4 Stimulus and Equipment

The ski map of the Alpes d'Huez resort (size A0, 119×84 cm) was presented in a wall-mounted backboard typical of the panel displayed at the bottom of the ski runs or at the start of ski lifts. Figure 6.3 shows a participant during the eye-tracking experience while wearing the eye movement–tracking device, the Tobii Pro glasses, which are advertised to unobtrusively capture natural viewing behavior in a real-world environment (www.tobiipro.com). The glasses track the pupil of the eye and sample gazes at the frequency of 50 Hz. Additionally, they are equipped with a wide-angle high-definition scene camera (90°).

6.6.1.5 Data Collection

Gaze data for each participant consisted of 3036 video frames for Task 1 and 2120 for Task 2. Over all participants, we collected 60 min of video-audio recordings.

FIGURE 6.3
Participant during the eye-tracking experience, wearing Tobii glasses.

6.6.1.6 Dependent Variables and Data Analysis

Dependent variables were fixation count and fixation duration, which were analyzed as a function of both participant proficiency level and the category of the fixated graphic elements. Gaze data were processed by Tobii Analyzer software with Custom I-VT filter. This filter refers to the Velocity-Threshold Identification (I-VT) fixation classification algorithm, which is a velocity-based classification algorithm (Salvucci & Goldberg, 2000). The general idea behind an I-VT filter is to classify eye movements based on the velocity of the directional shifts of the eye. The velocity is most commonly given in visual degrees per second. If the velocity falls above a certain threshold, the sample for which the velocity is calculated is classified as a saccade sample, and below the threshold, it is regarded as part of a fixation (Olsen & Matos, 2012). The Tobii Custom I-VT filter we applied had the main following settings: I-VT fixation classifier: 100 degrees/second and minimum fixation duration: 100 ms. Gaze data were analyzed with classical techniques:

Heatmaps: A heatmap uses different colors to illustrate the number of fixations participants made within some area of the stimulus, or for how long they fixated within that area. Red usually indicates the greatest number of fixations or the longest time and green the lowest, with varying levels color coded in between. Tobii Heat Map software calculates the distribution of color values around a fixation point by using an approximation to the Gaussian curve (cspline). The radius of this function is set by default at 50 pixels. Heat maps were rendered with absolute count and absolute duration of fixations.

Areas of interest (AOI): An AOI is a region of the stimulus that the researcher determines a priori to be relevant for the research question. If a fixation occurs in an AOI, it is generally assumed that the participant perceived the object encompassed by the AOI (Kiefer et al., 2014). Thirty AOIs were defined to analyze the task of exploration-valuation of the ski resort (Task 1). Each AOI includes one or more objects depicted in the map. Figure 6.4 shows AOI regions depicted on the ski map for Task 1. AOIs labeled G1–G4 and AOIs H-1 to H-3 encompass some large areas of the landscape that may be part of the "visit" to the resort as well as peaks on the horizon. The regions labeled P1–P8 enclose the main ski run areas, and the regions labeled R1–R10 enclose the main ski lift. A polygon labeled as the "Route" encompasses the main road down the valley, and the regions V1–V4 include the four main villages of the resort. According to the taxonomic categories of geographic objects described earlier, AOIs G and H refer to the Geomorphologic category, AOIs P and R refer to the Tracing category, and AOI V and "Route" refer to the Structure category.

FIGURE 6.4
Areas of interest depicted on the Alpes d'Huez ski map.

Twenty AOIs were identified as being directly pertinent to the task of creating a route (Task 2). The AOIS labeled as IGeo 1–4 encompass areas of the landscape that may be part of a route. The AOIs labeled as Ip 1–8 enclose the main ski run areas, and the AOIs labeled as Ir 1–10 encompass the main ski trail tracks. According to the taxonomic categories of geographic objects described earlier, the AOIs labeled as IGeo refer to the Geomorphologic category, while polygons Ip and Ir refer to the Tracing category. Figure 6.5 shows AOIs for Task 2.

6.6.2 Results

Heatmaps show that in the first 100 s of a ski area valuation task (Task 1), participants explore the central area of the map. Three patterns of eye movements corresponding to the ski level were revealed. Advanced skiers sweep the area within the boundaries of ski areas, with greater dwell along the longest, central line of ski lifts (which goes from the bottom to the top of the ski area). Intermediate-level skiers dwell along the longest, central line of ski lifts, which reaches the highest peaks. Novices' dwells are more limited to the very central area of the ski resort. Heat maps show a general lack of interest in horizons and panoramic borders of the map, and this held for all the experience levels. Figure 6.6 shows example heatmaps for Task 1.

In the wayfinding task (Task 2), the participants explored the main ski trails and trail connection points. Figure 6.7 shows example heatmaps for Task 2. Unlike the results from Task 1, the heatmaps for the three proficiency levels

FIGURE 6.5
AOIs for Task 2. The black line indicates the (ideal) route linking the starting and the arrival points.

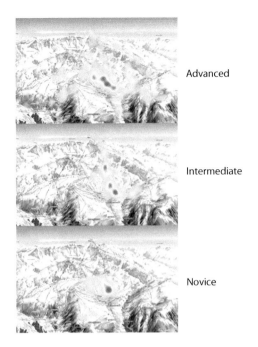

Advanced

Intermediate

Novice

FIGURE 6.6
Example of heatmaps for exploring and assessing the ski resort task (Task 1).

are rather similar. This is most likely due to the zones that are imposed by the route, but it is also noteworthy that participants at the three proficiency levels dwelled at the same places of the route. Participants' verbal annotations help us understand why (see Section 6.5.1.2).

The results address the second experimental hypothesis. Graphic objects representing ski trails and tracings of ski slopes are dwelled on the most during a ski area evaluation (Task 1) and in the wayfinding task (Task 2). The total fixation count for Task 1 shows that the objects of the Tracing category are gazed on by far the most. Figure 6.8 shows the total fixation count per object category and per participants' proficiency level.

Graphic objects representing the geomorphology and relief are not often the focus of dwells, so they seem not to be very relevant for the area evaluation. In Task 2, participants had to find and link two locations in the ski resort (the AOI and the black line indicating the "ideal" itinerary shown in Figure 6.5). Figure 6.9 shows the distribution of the total fixation count for Task 2 by participants' proficiency level.

Finally, combining gaze data and verbal notations, we can address the third experimental hypothesis and identify the "critical areas" of the map, those that generated misunderstanding, doubt, or uncertainty. In Task 2, they correspond to highly deformed areas with an important connection point to the ski network (crossing points and start and arrival points). Deformed areas are zones that have been distorted by the designer

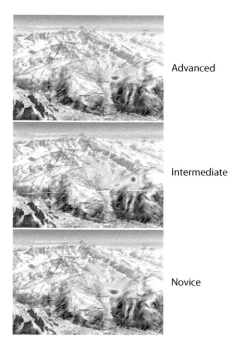

FIGURE 6.7
Example heatmaps for the wayfinding task (Task 2).

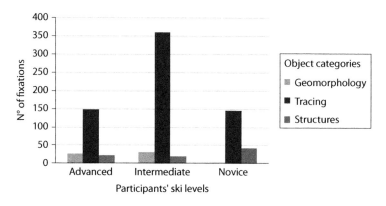

FIGURE 6.8
Total fixation count for graphic objects gazed at during the visual exploration of the ski map (Task 1).

(compressed, turned, or reduced) to satisfy the aesthetic requirements of panoramic drawing. These areas seem to attract the attention of the participants. A significant example is the crossing area from the central site of Alpes d'Huez region to the Auris en Oisans region, which is part of the route asked for in Task 2. Figure 6.10 is an aerial image showing the position of the two critical locations.

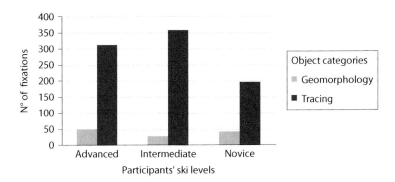

FIGURE 6.9
Total fixation counts for graphic objects gazed at during a creating route task (Task 2).

FIGURE 6.10
Actual position of Auris en Oisans and Alpes d'Huez. (Reproduced with permission from Google Maps images ©2015.)

To show Auris village in the map representation, the panoramist deformed the terrain: he tightened the valley and rotated the position of Auris village. Figure 6.11 shows the two locations in the ski map representation (surrounded by the red circles). It also shows an example of the heatmaps of the crossing area between the two locations. This crossing area at the bottom of the valley attracted a great dealt of visual attention from all of the participants.

Associated with the gaze pattern shown in Figure 6.11 were the following participant comments expressing difficulties in understanding the depicted information:

> "I can go by here, but it looks a bit confusing, I do not know why, maybe it's closeness to the slopes, the mountain ..."
> "This passage troubles me because I do not understand with this shadow, if the only way to cross is that track ... I do not know in which direction goes this curve. I hesitate to venture out there."

FIGURE 6.11
Example of a heatmap from the crossing area between Alpes d'Huez and Auris en Oisans.

To summarize, we identified the cartographic objects that are used by skiers to perform skiing tasks and map features or design aspects that posed problems of map interpretation. To see whether these findings might be generalized, we conducted an online survey.

6.7 Experiment 3

6.7.1 Online Survey

In our third study, we conducted a survey to gather opinions with a large number of participants ($n > 100$) and thus attempt to confirm previous results and justify their generalization. The survey inquired about three main topics: (1) During a skiing day, in which tasks is the ski map most used? (2) At what times of day are ski maps most used? and (3) What depicted information is more useful on a map? To answer these questions, a picture of the ski map of Alpes d'Huez was provided to the participant to accompany the survey.

Additional questions were designed to establish the participants' profiles: (1) knowledge about Alpes d'Huez, (2) level of skiing experience, (3) knowledge about cartography, and (4) demographic information.

The survey was administered online over a 15-day period. Half of the respondents were men and half women. About 67% of the participants were experienced skiers. Fifty-seven percent of the participants responded that they did not know the Alpes d'Huez ski resort.

6.7.2 Results

The results show that maps are relied on by 83% of participants. Most participants regarded maps as very useful in determining the starting point (67%), and at the other extreme, the lowest number of participants reported that maps are useful in determining ski trail run times. Figure 6.12 presents the percentages of participants who reported that maps are "very useful" for each of a number of tasks.

Paper ski maps are used mostly when skiers arrive at the resort, while they are taking a break and then need to get back to skiing. Paper maps do not seem to be used before coming to the ski resort. Typically, on the eve of a ski day, participants said they would visit the ski resort's website, which shows the resort's ski map. Figure 6.13 shows the percentages of responses according to the time within a skiing day.

During a typical skiing day, the most useful information is related to the Tracing category, followed by that from the Geomorphology category and that from the Structure category. A more detailed analysis of Tracings information, shown in Figure 6.14, shows that ski trail network and colored ski runs are the most important (response percentages greater than 75%).

Information from the Structures category seems not to be very important, with response percentages lower than 30%. However, Pictograms scored 31% and Place names 27.8%. Two kinds of geomorphological information were regarded as more important than the others: Elevation value and Peaks (response percentages greater than 40%). A detailed breakdown is shown in Figure 6.15.

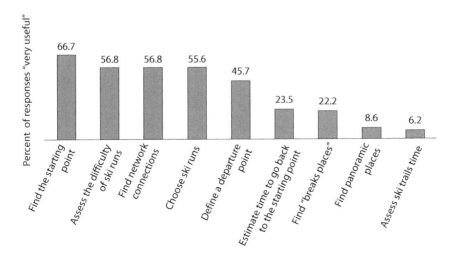

FIGURE 6.12
Ski maps' utility level for accomplishing a task in a ski resort.

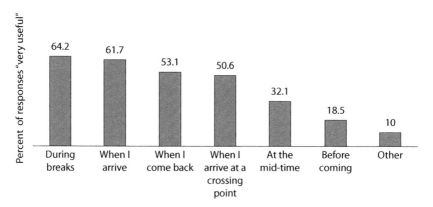

FIGURE 6.13
Times during a skiing day when maps are said to be very useful.

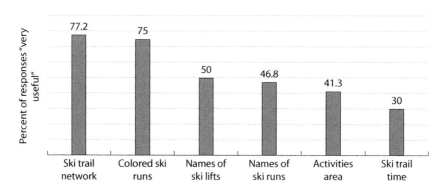

FIGURE 6.14
Very useful information from Tracing category.

These results confirm the insights overviewed in the analysis of verbal data and gaze data about the information that is used by skiers to make decisions.

6.8 Discussion and Outlook

Our experiments using protocol analysis, eye-movement tracking, and survey techniques allowed us to identify the geographic information (and its representation) that makes paper ski maps useful for skiers' cognitive tasks as well as geographic information that seems to be difficult for skiers to interpret.

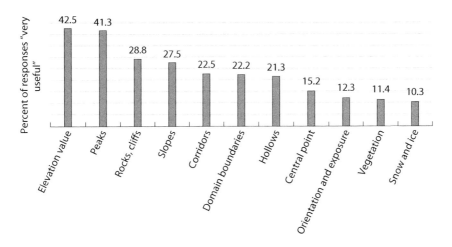

FIGURE 6.15
Geomorphology information rated as very useful.

For ski area valuation and path definition tasks (Experiments 1 and 2), the most helpful information seems to be that referring to mainly ski lifts, ski runs, and their connection points. Ski lifts and runs located in the focal point of the map, which tends to correspond to the central area of the resort and thus falls near the central region of the map, are those that attract the most attention. Skiers tend to focus their attention on the main (linear) structuring elements on the map. This finding is in agreement with the work by Ooms, De Maeyer, Fack, Van Assche, and Witlox (2012) on the design of complex maps. The authors indicate that to be able to create more effective maps, it is essential to represent data in a structured way. Also, linear geographic objects, such as rivers and road networks, are the preferred information suggested by users in the survey carried out by Petrovič and Masera (2005) concerning 3D map content.

Our results show that the geomorphic information and its graphic representation of landscape and landforms seem to pose more difficulties in interpretation, notably where the land surface has been deformed so as to make for a better panoramic. Analysis of the participants' explanations and of their visual attention during Tasks 1 and 2 in the experiments reveals that troubles in interpretation seem to be more evident in locations where there has been significant perspective-related distortion of reality. The concepts of local distortion and panorama view are intrinsically linked: the first is essential to create the second. Local distortions are unreal and are necessarily invented by the artist. They are made by rotation, exaggeration, reduction, and replication of shapes, and enlargement. The resulting graphical objects may be invisible in the actual scene, but that depends critically on the viewer's perspective (Balzarini et al., 2015).

According to the results of our study, a new research and map design question arises: Why continue to represent panorama views and the distortion of the terrain that they generate if this causes problems in the users' comprehension? One could respond in terms of uses and consider that ski maps should be based on different designs depending on the proposed use by the skier: to locate, to search, to appreciate a landscape, and so on. In fact, the ski map was originally created as a support for contemplation and subsequently became a support for decision making. In this sense, the panoramic view is a legacy of over 40 years of ski mapping, and it has influenced the landscape reading of generations of mountaineers, skiers, and resort operators. In awareness of this heritage, the answer is surely complex and deserves systematic study. The panoramic view still fascinates, and in this respect, the cultural, aesthetic, and emotional roles of panoramas need to be evaluated (Balzarini et al., 2015).

Immersive environments have started to replace traditional panoramic ski resort maps. With the latest generation of 3D ski maps (i.e., Fatmap apps [https://fatmap.com/], 4riders Social Ski map apps [www.4riders.ski/]), one can see the mountain area views represented at different scales, providing additional functional information. However, cognitive research on the impact that these technologies have on skier perception and decision making needs to be conducted. This will involve experiments on visual matching processes between objects in the natural environment and objects in the virtual environment display.

Effective visual exploration can be identified by analyzing the switches between real landmarks (i.e., peaks, rocks, and cliffs) and useful map symbols. Experiments in real conditions (i.e., at ski resorts) compared with laboratory experiments would also be worthwhile to investigate the balance between ecological validity and controllability (Kiefer et al., 2017).

Our experimental approach has some limitations, the main one being the relatively small sample size and thus, the relatively small number of participants in each of the three proficiency-level categories. Another important limitation was the time allotted to participants to perform their tasks (approximately 100 s for each task). This was necessary to facilitate image processing, which would otherwise have been too resource intensive. Another limit imposed by the eye-tracking method experience was reducing the original five categories of cartographic objects to three categories of AOIs. This grouping was done because of the difficulty of isolating gaze data collection on objects or AOIs with a very small area, such as pictograms or proper names.

Our results may not extend to map interpretation and decision making on longer ski map analyses, or indeed, on ski map interpretation and analysis conducted by pairs or small groups of skiers, which is normally the case in actual ski resort situations.

Our studies made it possible to validate experimental methods and the quality of the collected data. They therefore set the stage for more in-depth and real-world studies. A main contribution of our approach is in terms of methodology, which can provide relevant tools for further analysis of digital ski resort mapping and for innovative geovisualization design according to new practices in mountains. Our approach can be extended to other issues of effective design of geographic information for mountains (and not only) in the digital era.

Acknowledgments

This work has been supported by national grants LABEX ITEM ANR-10-LABX-50-01 and ANR-11-EQPX-0002. Access to the facility of the MSH-Alpes SCREEN platform for conducting the research is gratefully acknowledged. Grateful thanks to Arthur Novat for the time and enthusiasm he has devoted to the project and to Joanne Poulenard for her meticulous transcription of the large corpus of verbal data.

References

Balzarini, R., Dalmasso, A., & Murat, M. (2015). A study on mental representations for realistic visualization. The particular case of ski trail mapping. *ISPRS Archives* Vol. XL-3/W3, 495–502.

Belluard, N., Novat, A, & Novat, F. (2013). *Plan des pistes. Les domaines skiables de France dessinés par Pierre Novat.* Collection: Beaux Livres, Edition Glénat, Grenoble novembre 2013, EANI/ISBN: 9782723495431.

Bratkova, M., Shirley, P., & Thompson, W. B. (2009). Artistic rendering of mountainous terrain. *ACM Transactions on Graphics*, 28(4), 1–17.

Chi, T. H. M. (1997). Quantifying qualitative analyses of verbal data: A practical guide. *The Journal of the Learning Sciences*, 6(3), 271–315.

Çöltekin, A., Fabrikant, S. I., & Lacayo, M. (2010). Exploring the efficiency of users' visual analytics strategies based on sequence analysis of eye movement recordings. *International Journal of Geographical Information Systems*, 24(10), 1559–1575.

Crandall, B., Klein, G., & Hoffman, R. R. (2006). *Working Minds: A Practitioner's Guide to Cognitive Task Analysis.* Cambridge, MA: MIT Press.

Duchowski, A. T. (2007). *Eye Tracking Methodology: Theory and Practice* (2nd ed.). London: Springer.

Ericsson, K. A. (2006). Protocol analysis and expert thought: Concurrent verbalizations of thinking during experts' performance on representative tasks. In K. A. Ericsson, N. Charness, P. J. Feltovich & R. R. Hoffman (Eds.), *The Cambridge Handbook of Expertise and Expert Performance* (pp. 223–241). New York: Cambridge University Press.

Gentner, D., & Stevens, A. (Eds.) (1983). *Mental Models*. Mahwah, NJ: Erlbaum.

Hegarty, M., Montello, D. R., Richardson, A. E., Ishikawa, T., & Lovelace, K. (2006). Spatial abilities at different scales: Individual differences in aptitude-test performance and spatial-layout learning. *Intelligence*, 34(2), 151–176.

Holmqvist, K., Nyström, M., Andersson, R., Dewhurst, R., Jarodzka, H., & Van De Weijer, J. (2011). *Eye Tracking—Comprehensive Guide to Methods and Measures*. New York: Oxford University Press.

Jenny, H., Jenny, B., Cartwright, W. E., & Hurni, L. (2011). Interactive local terrain deformation inspired by hand-painted panoramas. *The Cartographic Journal*, 48(1), 11–20.

Kiefer, P., Giannopoulos, I., & Raubal, M. (2014). Where am I? Investigating map matching during self-localization with mobile eye tracking in an urban environment. *Transactions in GIS*, 18(5), 660–686.

Kiefer, P., Giannopoulos, I., Raubal, M., & Duchowski, A. T. (2017). Eye tracking for spatial research: Cognition, computation, challenges. *Spatial Cognition and Computation*, 17(1–2), 1–19.

Libarkin, J. C., & Kurdziel, J. P. (2002). Research methodologies in science education: The qualitative-quantitative debate. *Journal of Geoscience Education*, 50(1), 78–86.

Lloyd, R. (1993). Cognitive processes and cartographic maps. In T. Garling & R. G. Golledge (Eds.), *Behavior and Environment: Psychological and Geographical Approaches* (pp. 141–169). Amsterdam: North-Holland.

MacEachren, A. M. (1995). *How Maps Work: Representation Visualization and Design*. New York: Guilford.

Montello, D. R. (2002). Cognitive map-design research in the twentieth century: Theoretical and empirical approaches. *Cartography and Geographic Information Science*, 29(3), 283–304.

Montello, D. R. (2009). Cognitive research in GIScience: Recent achievements and future prospects. *Geography Compass*, 3/5, 1824–1840.

Newell, A., & Simon, H. A. (1972). *Human problem solving*. Englewood Cliffs, NJ: Prentice Hall.

Olsen, A., & Matos, R. (2012). Identifying parameter values for an I-VT fixation filter suitable for handling data sampled with various sampling frequencies. *ETRA '12*. Santa Barbara, CA. New York: ACM.

Ooms, K. (2016). Cartographic user research in the 21st century: Mixing and interacting. In T. Bandrova & M. Konečný (Eds.), *Proceedings of 6th International Conference on Cartography and GIS*, June 13, 2016, Albena, Bulgaria. ISSN: 1314-0604.

Ooms, K., De Maeyer, P., Fack, V., Van Assche, E., & Witlox, F. (2012). Interpreting maps through the eyes of expert and novice users. *International Journal of Geographical Information Science*, 26(10), 1773–1788.

Patterson, T. (2000). A view from on high: Heinrich Berann's panoramas and landscape visualization techniques for the U.S. National Park Service, *Cartographic Perspectives*, 36.

Petrovič, D., & Masera, P. (2005). Analysis of user's response on 3D cartographic representations. In *Proceedings of the 22nd ICA International Cartographic Conference*, Coruna, Spain.

Salvucci, D. D., & Goldberg, J. H. (2000). Identifying fixations and saccades in eye-tracking protocols. *Proceedings of the Symposium on Eye Tracking Research & Applications, ETRA '00*. Palm Beach Gardens, FL. New York: ACM, pp. 71–78.

Spengler, M., & Räber, S. (2012). Panoramic maps—evaluating the usability and effectiveness. *Proceedings of 8th ICA Mountain Cartography Workshop.* Taurewa, New Zealand. Auckland, New Zealand: Cartopress, New Zealand Cartographic Society, Inc.

Tait, A. (2010). Mountain ski maps of North America: Preliminary survey and analysis of style. *Cartographic Perspectives*, 67, 5–18.

Ware, C. (2008). *Visual Thinking for Design.* Burlington, MA: Elsevier.

7

The Human Factors of Geospatial Intelligence

Laura D. Strater, Susan P. Coster, Dennis Bellafiore, Stephen P. Handwerk, Gregory Thomas, and Todd S. Bacastow

CONTENTS

KEYWORDS: *geospatial intelligence, human factors, goal directed task analysis (GDTA), sensemaking, GEOINT*

7.1 Introduction

The goal of this chapter is to better understand the human factors (HF) challenges associated with geospatial intelligence. HF engineering is a multidisciplinary field that studies interactions between humans and technology to improve overall human-system performance while promoting the good of the human. Formally, the International Ergonomics Association defines human factors as "the scientific discipline concerned with the understanding of interactions among humans and other elements of a system … in order to optimize human well-being and overall system performance" (International Ergonomics Association, 2017). Alphonse Chapanis (1991), one of the pioneers of human factors, defines it as:

> A body of knowledge about human abilities, human limitations, and other human characteristics that are relevant to design. Human factors engineering is the application of human factors information to the design of tools, machines, systems, tasks, jobs, and environments for safe, comfortable, and effective human use (p. 1).

This definition extends the concept of human factors beyond the design of systems to the study and design of ways of improving "cognitive work": to understand how emerging technologies and learning methods may best serve the geospatial intelligence analyst's cognitively intensive work in a complex human-technical "ecosystem."

This chapter discusses the cognitive basis of geospatial intelligence by considering the operator goals—*why* the analyst is completing tasks rather than *how* the analyst is completing tasks. We elaborate on the tradeoff between goal-driven and data-driven processing using the Data/Frame model of sensemaking (Klein, Moon, & Hofman, 2006a,b; Sieck et al., 2007). Further, this chapter examines underlying human factors that have the potential to amplify and extend the geospatial analyst's abilities by focusing specifically on cognitive processes. We organize our understanding of these cognitive processes with a preliminary cognitive task analysis, employing the goal directed task analysis method (GDTA) (Endsley, 1993), to examine the critical cognitive human factors challenges in geospatial intelligence. Our findings provide insight into the potential for promoting the capabilities of the geospatial analyst through education, training, and advanced system design that carefully considers human abilities and limitations.

7.2 Geospatial Intelligence and the Automation Bias

The U.S. government defines *geospatial intelligence* in U.S. Code Title 10, §467 (2006) as follows:

> The term "geospatial intelligence" means the exploitation and analysis of imagery and geospatial information to describe, assess, and visually depict physical features and geographically referenced activities on the earth. Geospatial intelligence consists of imagery, imagery intelligence, and geospatial information.

The legal definition was written to form the National Geospatial Intelligence Agency (NGA). A more applicable definition for this chapter has emerged, which describes geospatial intelligence as a field of knowledge, a process, and a profession (Bacastow & Bellafiore, 2009). As knowledge, it is information integrated in a coherent space-time context that supports descriptions, explanations, or forecasts of human activities using which decision makers take action. As a process, it is the means by which data and information are collected, manipulated, geospatially reasoned, and disseminated to decision makers. The geospatial intelligence profession establishes the scope of activities, interdisciplinary associations, competencies, and standards in academe, government, and the private sectors.

The intelligence community in general, and the geospatial intelligence community specifically, is overwhelmed with data (Young, 2013). There is evidence that this flood of data increases the analyst's uncertainty (Klein, 2015). This causes a particular problem for the geospatial intelligence analyst. The collection of imagery, imagery intelligence, and geospatial information—the legally defined elements of geospatial intelligence—is the most easily and often automated component of the intelligence cycle. Thus, while the "big data" glut is problematic for intelligence in general, some of the most extreme problems are seen in the geospatial intelligence arena. The prospect is that the data glut will grow and, as Klein suggests, increase analyst uncertainty, since systems are now collecting megabytes of data for each human on the earth every minute of the day (EMC, 2014).

There is a predisposition toward technical solutions in geospatial intelligence. The bias is driven by the impossibility of hiring, training, and supporting the army of human analysts required to manually process the immense volume of imagery and geospatial data (Frommelt, 2015). The expression *deus ex machina* has been used to describe an approach to geospatial intelligence related to the "too much data" and "not enough people" dilemma (Frommelt, 2015). Greek tragedies used a mechanical device, known as a *mechane*, to dramatically insert a character into a scene. The character, often a god, would use its power to resolve the play's problem and bring a quick solution. *Mechane* became so normal in Greek drama that the Latin phrase *deus ex machina* was coined to mean "god from the machine," but it came to mean an implausible but convenient solution to a complex situation. In geospatial intelligence, the expression describes the hope that technology can quickly and "divinely" solve a human dilemma—providing a solution to an analytic problem.

Geospatial intelligence analysis is ultimately about anticipating another's perception of events in space-time. The *mechane* only aids the human in developing analytic insight. The clear human factors challenge for geospatial intelligence analysis boils down to the following: to optimally integrate the human and machine functions to develop insights about how another human will use the landscape to their advantage.

Will geospatial intelligence follow in the footsteps of theater and provide implausible and incomprehensible solutions, or will technologists find a way to develop capable systems that can effectively and synergistically team with geospatial analysts? Only time can provide this answer. What is certain is that geospatial intelligence insights are developed within an ecosystem of humans and machines, where machines play a role in selecting and preparing information for analysts' consumption. HF concepts and methods have the potential to significantly improve overall human-system performance.

7.3 What Geospatial Intelligence Analysts Do

Our focus is on the essential and most resistant-to-change portion of the geospatial intelligence ecosystem—the human analyst. The following are presented as unifying concepts to appreciate the analyst's and the machine's roles in the context of the discipline (adapted from Bacastow, 2016):

- Geospatial intelligence seeks knowledge to achieve a decision advantage.
- Geospatial intelligence reveals how human behavior is influenced by the physical landscape, time, and human perceptions of the earth.
- Geospatial intelligence reshapes understanding by discovering relationships through space and time.

The human component of geospatial intelligence is ultimately where knowledge work is done and insights produced; thus, it is dependent on the geospatial analyst's know-how. Although often confused with a workflow, the analyst's cognitive actions are not a sequence of steps through which work passes; however, cognitive activities may be associated with parts of a workflow. The analyst's cognitive actions can generally be classified as *sensemaking*, a concerted cognitive effort to understand the relationships among disparate objects and events to place them within a context or frame that has explanatory power.

Geospatial analysis thinking begins with the conceptualization of space-time to frame the problem. In the geospatial sciences, this is called a *mental map* (Gould & White, 1993). The nature of the frame is critical, because it ultimately determines the interpretation of the analyst's observations—how they make sense of the geospatial relationships. These are generally considered within the three geographic frames—physical spaces, behavioral spaces, and cognitive spaces (National Academies Press, 2006, p. 28). The frame provides the interpretive context that gives meaning to the geospatial data. The three general geospatial frames are described as follows:

- *Physical space* is built on the four-dimensional world of space-time but focuses on the physical.
- *Behavioral space* is the four-dimensional space-time that focuses on the spatial relations and interactions between the individual actors and objects in the physical environment.
- *Cognitive space* focuses on concepts and objects that are not in and of themselves necessarily spatial, but the nature of the space is defined by the particular problem. This is the mental map that exists in the analyst's mind.

These conceptual frames, or maps, provide an interpretative context to explain why objects are where they are and how the analyst decodes relationships between and among the objects. The geospatial frames can be fixed into a workflow or analytic process. For example, NGA adopted an analytic workflow known as *geospatial-intelligence preparation of the environment* (GPE) (NGA, 2006) for military applications of geospatial intelligence. The first step is to *define the environment*. This involves collecting data on the area of interest (AOI) in terms of space-time that focuses on the physical, political, and ethnic boundaries. This step initiates sensemaking. Next, the analyst must *describe influences of the environment*, which elaborates on the AOI to consider factors that focus on the spatial relations and interactions between the individual actors and objects in the physical environment. The third step is to *assess threat and hazards*, which involves creating a frame of intent. The frame focuses on concepts and objects that are not in and of themselves necessarily spatial. This is the mental map that exists in the analyst's mind. Finally, the analyst synthesizes, or makes sense of, the frames to *develop analytic conclusions*.

An important point is that while we use this military example of GPE, this process extends across all types of geospatial intelligence analysis. An analyst supporting the search for a lost child in a national park will also conduct threat analysis, though they may not even be considering the threat of a hostile human actor but rather, the threats in the environment. This analyst will also conduct predictive analysis, using what is known about the terrain and typical human behavior to determine the outer limits of the search area and the most likely areas to find the lost child.

While these principles delineate the geospatial intelligence profession, they also teach us the geospatial aspects of the sensemaking process, where the human and technical components of geospatial intelligence analysis mix. Machines automate processes that increase the efficiency of the human performance of the analytic task. However, at its most basic core component, the data, systems, and humans work in concert. As such, this requires an understanding of the synergy between the human cognitive thought processes involved and the technical systems used.

7.4 Sensemaking

Cognitive research on intelligence analysis has highlighted that analysts engage in an iterative sensemaking process that involves considering the data available from different viewpoints and perspectives (c.f., Pirolli & Card, 2005; Klein et al., 2006b). Sensemaking has been defined as "a motivated, continuous effort to understand connections (which can be among people, places, and events) in order to anticipate their trajectories and act effectively" (Klein et al., 2006a, p. 71). Moreover, Klein et al. (2006b) posit a generic Data/Frame theory of sensemaking, which suggests that analysts collect sufficient data to establish an initial frame (or mental map) for making sense of the data. The data used to create the frame are "the interpreted signals of events," and frames are "the explanatory structures that account for data" (Klein, Phillips, & Peluso, 2007, p. 120). This frame can be considered an organizing entity—a specific instance of a mental model of the elements and relationships under analysis.

Sieck et al. (2007) as well as Klein et al. (2006) have described six actions of sensemaking:

- Seeking a frame
- Questioning the frame
- Elaborating the frame
- Preserving the frame
- Comparing frames
- Reframing

One of the challenges in the study of intelligence analysts of any type is that much of their work is sensitive or classified. Thus, there are limited publicly available resources to help people understand the cognitive work of the analyst. To help address this problem and to focus our own research efforts, we conducted a preliminary cognitive task analysis using an abbreviated GDTA methodology. The output of the cognitive task analysis is a goal hierarchy describing the analyst's goals and subgoals as s/he seeks to understand the situation, which creates a framework for understanding the cognitive work of the geospatial analyst. One important note is that geospatial intelligence analysts actually work in many intelligence arenas. Geospatial intelligence analysts may work with any branch of the Department of Defense, with the intelligence community, or with law enforcement, for example. Although the problems they analyze may vary across domains, at some level of abstraction, the goals and decisions, and of course the analytic methods, are common across these domains. We chose to focus on the commonalities across domains rather than the differences.

Next, we describe our cognitive task analysis method and results.

7.5 Geospatial Intelligence Analyst Goal Directed Task Analysis

Our primary method for studying geospatial intelligence analyst cognition is a preliminary cognitive task analysis using a modified GDTA methodology (Endsley, 1993), based on hierarchical task analysis (HTA) (Annett & Duncan, 1967), conducted with a small number of geospatial intelligence analyst subject matter experts (SMEs). The GDTA focuses on user, or analyst, goals, on the decisions, assumptions, assessment, and judgments that analysts must make to successfully accomplish these goals, and on the heuristics they employ to understand the data. Cognitive task analysis often explores the things that make the analyst's job particularly demanding and difficult rather than on the more mundane task elements. The overall objective of a GDTA is to understand how experts think about solving a problem and determine what information they require when doing so and to gain insight into how they make decisions and resolve inconsistencies in the data. A typical GDTA, like most cognitive work analysis, begins with the review of any available materials that describe how the analysts perform their work, but the primary input to the GDTA comes from direct interviews with individual SMEs. Typically, each interview is conducted with a single expert and lasts approximately two hours. The GDTA seeks to identify the major goals of the user along with any subgoals for meeting each higher goal, the decisions associated with each subgoal, and the specific bits of information that the analyst needs to make the decision and meet the subgoal. After several interviews have been conducted, the researcher begins to analyze the data, organizing the information gathered into a goal hierarchy. The GDTA analysis then identifies the critical decisions that the user makes to meet each subgoal. For each decision, the information the user would ideally like to know to make the decision is delineated. Once the initial goal hierarchy has been completed, researchers follow up with additional expert interviews to fill in gaps in the structure. There is also a process of review, in which either a small number of the most informative and expert SMEs are selected to review the goal hierarchy or additional, highly expert SMEs are brought in for the review.

Our method varied from this standard GDTA methodology in several key ways. First, we were limited to only four experts who were available, and although our SMEs are considered experts in the field of geospatial intelligence, they are primarily geospatial intelligence instructors rather than geospatial intelligence analysts. See Table 7.1 for details on the qualifications of our experts. In addition, since we were looking for commonalities across geospatial intelligence applications, we were less focused on the specific information requirements our experts needed to make their decisions, since that can vary considerably across domains, and more on their processes and the types of information they used. Analysts will often say that they do not

TABLE 7.1

Subject Matter Experts Interviewed for the GDTA

Expert	Years as Analyst	Years as Instructor	Domains of Expertise (e.g., IC, DoD, Law Enforcement)
1	26	13	Law enforcement
2	6	30	DoD
3	3	35	IC
4	27	10	Business

Note: IC, U.S. Intelligence Community; DoD, U.S. Department of Defense.

make recommendations, although they do make assessments based on the data, which constitutes their expert opinion and related recommendations. A sampling of the questions asked during the interview sessions is shown in Table 7.2. Moreover, our initial interview session was a group interview with three of our geospatial intelligence analysis SMEs. While this is generally not ideal, the interviewers made a conscious effort to seek input from all participants, including dissenting views. SMEs were told that we were not looking for agreement at this stage; rather, we were looking for diversity of opinion. This was followed by a more typical interview session with one additional SME. After this second session, the goal hierarchy was created and was then reviewed in two additional group sessions with all four SMEs.

The data from the GDTA interviews were incorporated into the high-level goal hierarchy, shown in Figure 7.1. The overall goal for a geospatial intelligence analyst is to *use geospatial analysis methods to "find truth" in response to requests for information.* We recognize that the goal of finding truth is not always achievable, particularly in any area of intelligence analysis, where there is often an opponent seeking to hide that truth from view. We believe, however, that finding truth is still the high-level goal, despite the difficulty in attainment. At its core, finding truth is the goal of all sensemaking, although the operational reality may be that getting close enough to truth to defeat the opponent is sufficient. Sensemaking is the cognitive process of iteratively fitting data to a frame and fitting a frame to the data (Klein et al., 2006a,b; Sieck et al., 2007). The data-frame model of sensemaking provides a description of

TABLE 7.2

Example Interview Questions

Goal Queries	Assessment Queries	Information Queries
What are the primary goals of your job as a geospatial intelligence analyst?	What are some of the most critical decisions or assessments that a geospatial intelligence analyst must make?	What information do you need to make that decision?
What is your highest-level goal?	What assessments or decisions are you making to achieve that goal?	How will you use that information in making your decision or assessment?
When trying to meet this goal, what subgoals or objectives do you try to meet?	What determines when you have sufficient information for an assessment or recommendation?	How will you combine multiple pieces of information into an assessment? Which pieces go together?
What triggers make you shift from one goal or subgoal to another?	What would trigger you to reevaluate your assessment?	What are the projections of future state that you're trying to make?
What goals are specifically related to the geospatial aspects of the data?	Describe some methods for evaluating the geospatial relationships of the data.	What indicators are you looking for that provide information about intent?

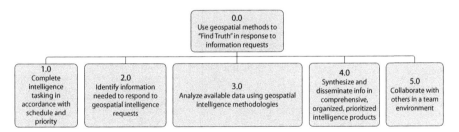

FIGURE 7.1
High-level goal hierarchy for geospatial intelligence analyst.

how the geospatial analyst works at the most rudimentary level. The model describes how people construct and revise internal mental structures when they make sense of events, and the goal of this sensemaking process is to find truth by selecting the right frame to interpret the data.

A more detailed goal hierarchy is shown in Figure 7.2. The first subgoal identified from the GDTA interviews is to *complete intelligence tasking in accordance with schedule and priority*. This involves finding the answer to a geospatial intelligence question. The tasking will include the requirements for schedule, and priority is included in the tasking requirements. This indicates that while the goal is to complete all tasking on schedule, higher-priority tasks may bump lower-priority tasks. Again, sensemaking is always done for a purpose, and this goal reflects the reality that geospatial intelligence professionals are often tasked with multiple requests, and balancing schedule and priority across multiple competing demands requires effort. To achieve this goal, the analyst also has three subgoals: *evaluate geospatial collection requirements*, *evaluate the best sources for needed geospatial intelligence data*, and *analyze how and why locations of objects of interest change over time*. In these subgoals, the analyst will determine the requirements of the collection tasking and the best opportunities and options for collecting the data, and will begin the process of geospatial analysis by investigating the geospatial relationship among the objects of interest, looking at activity over time.

Next, analysts strive to *identify information needed to respond to requests* for geospatial intelligence information. All the steps of the GPE can be involved here, but the particular emphasis is on the three activities of defining the area, understanding the relationships between elements in the area, and assessing threats and hazards in the area, which are all information that will be needed to respond to requests. This is analogous to the *representing the situation* facet of sensemaking, the first facet, in which the sensemaker tries to pull relevant data from the stream of information while discarding data that are deemed not applicable to the current problem. The related subgoals are *identify relevant available geospatial intelligence data*, *evaluate available geospatial intelligence data for quality, timeliness and applicability*, *identify gaps in current data*, and *request information to fill gaps in the data*. To achieve these goals and effectively represent the situation, the analyst will determine what

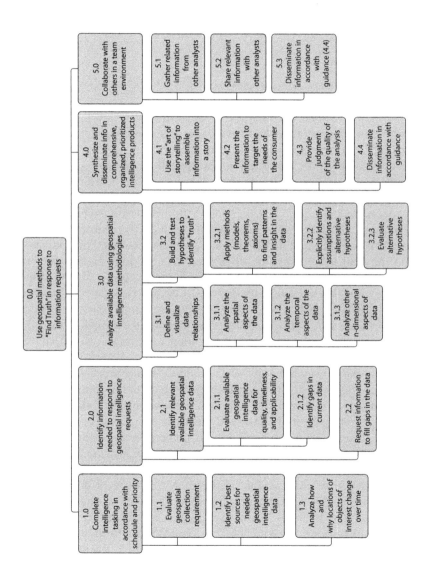

FIGURE 7.2
Goal hierarchy for geospatial intelligence analyst showing both goals and subgoals.

information is available, analyze the quality of these data, identify critical gaps in information needed to meet the tasking, and then determine how to fill those data gaps.

Our goal hierarchy next shows that analysts have a subgoal to *analyze the available data using geospatial intelligence methodologies*. This includes considering and presenting the geospatial and temporal data relationships and trying to extract meaning from these relationships. In the sensemaking model, this is when the analyst examines the relationships in the data to find an initial frame that fits the relationships and then elaborates and questions the frame. Data that have been discarded may be considered again as the analyst questions the frame to determine whether another frame can better explain the geospatial relationships among the data. In describing the importance of proximity in geospatial analysis, one of our SMEs paraphrased Tobler's first law of geography: "All things are related, but those that are closer are more related." Although proximity is an important concept in geospatial intelligence, other relationships are important as well. As a subgoal, the analyst will *define and visualize data relationships*, including analyzing the spatial, temporal, and multidimensional aspects of the data. This analysis of relationships assists the analyst in understanding the data to select a frame.

Seeking a frame is characterized by an attempt to understand the current situation, using the data obtained through the GPE process, to help the analyst understand the geospatial relationships between the actors and entities within the AOI. Through a process of pattern matching to familiar frames, the analyst selects a frame and then questions whether the frame really fits the current data. This may involve seeking additional information, or seeking to better understand how the elements are related, as the analyst elaborates the frame and documents the current frame to preserve it. The analyst has a goal to *build and test hypotheses to identify "truth."* During the process of elaborating and questioning the frame, the analyst will apply a variety of models (which are often maps), theorems, and axioms to search for patterns and meaning (frames) in the data. In particular, geospatial intelligence analysts seek to gain a visual sense of the data. Imagery is a critical component of analysis. The analyst strives for reliability and precision in the data, as even small variances can affect the perceived relationships among data points. In sensemaking, this corresponds to questioning the frame and reframing when a different frame better explains the relationships among the data. In addition, the skilled analyst considers other possible frames, using a method such as analysis of competing hypotheses, in which the analyst attempts to either validate or nullify the current frame. Depending on the outcome, the analyst may select another frame (reframing) that better explains the geospatial relationships and activity that have been identified.

To complete the intelligence tasking, the analyst must *synthesize and disseminate information in comprehensive, organized prioritized intelligence products*. This requires that the analyst is satisfied—for the moment—with the selected sensemaking frame and is willing to present the final data, within a frame that offers explanatory value, as an analytic work product. The information is integrated within the final frame, and storytelling is used to create a product that explains all of the data within the selected frame in a way that targets the needs of the consumers of the report. The work product will include an assessment of the quality of the data and the resulting analysis and will be shared in accordance with their guidance. It is important to note that geospatial analysts often develop expertise in a specific geographic area and that the delivery of a report is never the end of the analysis. In fact, if additional information is uncovered that causes a shift in the frame, or adds additional explanatory value to the existing frame, the analyst will revise the report to capture the additional data.

In addition, the analyst has a goal to *collaborate with others in a team environment*. Collaboration is increasingly common within the geospatial intelligence community, with multiple analysts working together to provide an integrated product that is more complete than any one expert analyst can provide and that spans multiple analytic specialties. As the geospatial intelligence analyst identifies relevant information for the analysis, s/he will reach out to other analysts who specialize in signals intelligence, imagery intelligence, human intelligence, or another intelligence field. The geospatial intelligence analyst will also be called on to apply geospatial expertise to help these other analysts in making sense of their data and to share data with those who are authorized to receive it.

Collaboration in intelligence extends beyond human to human. As we discussed earlier in this chapter, technology is prevalent within geospatial intelligence, and the geospatial intelligence analyst uses a number of tools to better perform the analysis tasks. Many of these are of great benefit and have extended geospatial analyst capabilities tremendously, such as machine-aided object recognition within imagery. However, as tools become more automated, the challenge is to maintain the analyst's ability to determine the accuracy and reliability of the data provided. As systems become more autonomous, they become more independent entities, which can contribute more directly to team analysis and success—but they can also contribute to breakdowns in communication and may lead to longer, rather than shorter, decision cycles.

7.6 Discussion and Conclusion

In this chapter, we have presented an HTA that provides a framework for thinking about the cognitive tasks of a geospatial intelligence analyst. Applying the data-frame model of sensemaking to geospatial intelligence analysis can be said to be the process of fitting geospatial data into a frame and fitting a frame around the geospatial data. The geospatial analyst's frames are mental models or maps that account for the data. The analyst's mental model and the data work in concert to generate an explanation that contributes to a question that need not be exclusively, or even overtly, geospatial.

This suggests that the geospatial analyst does not just analyze data by dis-aggregation. Instead, the geospatial analyst interprets data by comparing observed patterns of details with what they understand. Something makes sense because the analyst has seen a similar pattern in the past, and the similarities between the two patterns help the analyst make non-obvious inferences and draw conclusions. The analyst may even employ a new, self-generated pattern based on previously learned and remembered patterns if the analyst does not get a good match to an ostensible pattern (Moore, 2011, p. 7). Thus, the analyst's background, knowledge, and experience are central to the ability to select appropriate frames.

As with most cognitively demanding work, the novice is at a distinct dis-advantage in comparison to the expert. The expert's rich experience and knowledge base give them a greater variety of familiar frames and greater insight into the more subtle cues that lead to the selection of the most effective frame. Since the selected frame helps allocate limited attentional resources by directing attention to the items that are most relevant based on that frame, this advantage for the expert is significant. This suggests that technology or training that can rapidly expose analysts to a variety of frames, or that aids analysts in either selecting or questioning a frame, could be very beneficial.

References

Annett, J., & Duncan, K. D. (1967). Task analysis and training design. *Occupational Psychology*, 41, 211–221.

Bacastow, T. (2016). Viewpoint: A call to identify first principles. (2016, Jan 27) NGA Pathfinder. Viewpoint: A call to identify first principles, Retrieved from https://medium.com/the-pathfinder/viewpoint-a-call-to-identify-first-principles-d5e21cb2ce40.

Bacastow, T. S., & Bellafiore, D. J. (2009). Redefining geospatial intelligence. *American Intelligence Journal*, 38–40.

Cacciabue, P. C., & Hollnagel, E. (1995). Simulation of cognition: Applications. In *Expertise and Technology: Cognition and Human–Computer Cooperation*, 55–73.

Center for the Study of Language and Information (U.S.), & Stanford University. (1997). Stanford encyclopedia of philosophy. Retrieved from https://plato.stanford.edu/entries/thought-experiment/.

Chaitin, G. J. (2005). *Meta Math: The Quest for Omega* (1st ed.). New York: Pantheon Books.

Chapanis, A. (1991). To communicate the human factors message, you have to know what the message is and how to communicate it. *Human Factors Society Bulletin*, 34(11), 1–4.

Cooper, R., & Foster, M. (1971). Sociotechnical systems. *American Psychologist*, 26, 467–474.

Coquillette, D. R. (1994). Professionalism: The deep theory. *North Carolina Law Review*, 72(5), 1271.

EMC. (2014). Executive summary: Data growth, Business opportunities, and the IT imperatives, April 2014, Digital Universe with Research & Analysis by IDC. Retrieved from www.emc.com/leadership/digital-universe/2014iview/executive-summary.htm.

Endsley, M. R. (1993). A survey of situation awareness requirements in air-to-air combat fighters. *International Journal of Aviation Psychology*, 3(2), 157–168.

Frommelt, P. (2015). Persistence: How we get and convey geospatial intelligence. Retrieved from www.nga.mil/MediaRoom/News/Pages/Persistence-PF.aspx

Gould, P., & White, R. (1993). *Mental Maps*. New York: Routledge.

Gregory, D., Johnston, R., & Pratt, G. (2009). *Dictionary of Human Geography: Mental Maps/Cognitive Maps* (5th ed.). Hoboken, NJ: Wiley-Blackwell. p. 455.

Hoffman, R. R., Shadbolt, N. R., Burton, A. M., & Klein, G. (1995). Eliciting knowledge from experts: A methodological analysis. *Organizational Behavior and Human Decision Processes*, 62(2), 129–158.

International Ergonomics Association. (July 12, 2017). Definition and domains of ergonomics. Retrieved from www.iea.cc/whats/index.html

Klein, G. (2015). A naturalistic decision making perspective on studying intuitive decision making. *Journal of Applied Research in Memory and Cognition*, 4(3), 164–168. 10.1016/j.jarmac.2015.07.001.

Klein, G., Moon, B., & Hoffman, R. R. (2006a). Making sense of sensemaking 1: Alternative perspectives. *IEEE Intelligent Systems*, 21(4), 70–73.

Klein, G., Moon, B., & Hoffman, R. R. (2006b). Making sense of sensemaking 2: A macrocognitive model. *IEEE Intelligent Systems*, 21(5), 88–92.

Klein, G., Phillips, J. K., & Peluso, D. A. (2007). A data-frame theory of sensemaking. In *Expertise Out of Context: Proceedings of the Sixth International Conference on Naturalistic Decision Making*.

Moore, D. T., & Center for Strategic Intelligence Research (U.S.). (2011). *Sensemaking: A Structure of an Intelligence Revolution*. Washington, DC: Center for Strategic Intelligence Research, National Defense Intelligence College.

National Academies Press (U.S.), & ebrary, I. (2006). *Learning to Think Spatially*. Washington, DC: National Academies Press.

NGA. (2006). *National System for Geospatial Intelligence: Geospatial Intelligence (GEOINT) Basic Doctrine*. Office of Geospatial-Intelligence Management.

NGA. (2017). NGA Products and services: GEOINT analysis. Retrieved from www.nga.mil/ProductsServices/GEOINTAnalysis/Pages/default.aspx

Pirolli, P., & Card, S. (2005). The sensemaking process and leverage points for analyst technology as identified through cognitive task analysis. In *Proceedings of International Conference on Intelligence Analysis (Vol. 5)*.

Sieck, W. R., Klein, G., Peluso, D. A., Smith, J. L., Harris-Thompson, D., Gade, P. A., & Klein Associates Inc., Fairborn, OH. (2007). FOCUS: A model of sensemaking. US Air Force Technical Report 1200

United States Code. (2006). Title 10 - Armed Forces, Sec. 467 – Definitions

Young, A. (2013). Too much information: Ineffective intelligence collection. *Harvard International Review*, 35(1), 24–27.

8

Employing Ontology to Capture Expert Intelligence within GEOBIA: Automation of the Interpretation Process

Sachit Rajbhandari, Jagannath Aryal, Jon Osborn,
Arko Lucieer, and Robert Musk

CONTENTS

KEYWORDS: *GEOBIA, ontology, human factor, rule-based classification*

8.1 Introduction

The importance of remote sensing image analysis is ever increasing due to its ability to supply meaningful geographic information that informs local and global problems, such as measuring urban sprawl, mapping vegetation communities, monitoring the impacts of global climate change, and managing natural resources and urban planning. In this process of geo-object extraction, geographic object-based image analysis (GEOBIA) provides a method to identify real-world geographic objects from remotely sensed imagery. GEOBIA uses techniques analogous to those used by humans to perceive and distinguish geo-objects in imagery, usually acquired from satellite or airborne platforms. Experts use domain knowledge and measurement data

extracted from remote sensing images for object-based analysis. This signi-
fies a need for human involvement in the form of applying expert knowledge
at the time of image object identification. The need for such human intervention acts as a barrier to the automation of GEOBIA processes. In this regard,
knowledge representation techniques such as the use of ontologies provide
possibilities for modeling expert knowledge in a manner that contributes
to the further development of GEOBIA. In this chapter, we will discuss the
importance of the human factors in GEOBIA. To this end, we will draw on
literature from both GEOBIA and ontology use.

GEOBIA has become a new paradigm for remote sensing image analysis
because of its capability to incorporate spectral, spatial, temporal, contextual, and
in particular, human experience and expertise in image analysis (Blaschke et al.,
2014). GEOBIA has emerged as a new paradigm to replicate the human interpretation of images in a semi-automated way for accurate geo-object identification,
with reduced time and cost of human involvement (Hay & Castilla, 2008).

Before GEOBIA, per-pixel methods were predominantly used. In these
methods, single pixels are assigned to different geo-object classes. The "per-
pixel approach" works at the spatial scale of a pixel. There will be situations
when a single pixel may represent more than one object if the size of the
objects is smaller than the pixel size. Similarly, if the size of the objects is
larger than a single pixel, we need more than one pixel to represent the object
of interest (Aplin & Smith, 2008; Blaschke, 2010). Thus, with an increase in
the spatial resolution of satellite images, per-pixel classification suffers from
the so-called "salt and pepper" effect, whereby a single pixel may be mis-
classified as a different class within a homogeneous region (Blaschke, Lang,
Lorup, Strobl, & Zeil, 2000; Weih & Riggan, 2010). In high-resolution images,
one is more likely to find that neighboring pixels belong to the same object
class, allowing the grouping of homogeneous pixels for image object delineation and classification (Blaschke & Strobl, 2001). A further limitation of the
traditional per-pixel analysis is that it only makes use of spectral information without considering spatial or contextual information that could make a
significant contribution to image analysis (Weih & Riggan, 2010).

Given these shortcomings of per-pixel methods, the remote sensing community recognized the need for an object-based approach such as GEOBIA.
Advances in satellite imaging technology are making very-high-resolution
remotely sensed image data increasingly available. For example, spatial resolution has improved from 80 m pixels in Landsat 1 images (Lauer, Morain, &
Salomonson, 1997) to 1.24 m in WorldView-3 images (DigitalGlobe, 2014). In
such scenarios, image analysis at the pixel level becomes impractical when
there is a need to process and extract information from a large number of
high-resolution satellite images. Also, there exists a "semantic gap" between
the object of interest and image data. With high-resolution images, one pixel
cannot correspond to one semantic object. Thus, to represent a geo-object, a
group of homogeneous contiguous pixels needs to be combined. When one

is dealing with a single pixel, the surrounding pixels may not provide useful information. However, in GEOBIA, a neighboring group of pixels representing a geo-object provides significant contextual information that may be useful for image object identification. For example, a group of pixels can be classified as a lake if it represents a water object surrounded by land. GEOBIA makes use of contextual information in its object identification task.

The key benefit of using GEOBIA is the capacity to incorporate expert knowledge during the image segmentation and classification process. Knowledge about domain area, sensor characteristics, appropriate image scale, and contextual dependencies assists in extracting information that is useful for the image interpretation process (Benz, Hofmann, Willhauck, Lingenfelder, & Heynen, 2004). The knowledge may be available in the form of an expert's domain knowledge, image analysis experience, or familiarity with relevant literature (Argyridis & Argialas, 2015). Human interaction is required to process such knowledge while introducing minimal or no bias, which limits the capacity of GEOBIA to be fully automated (Belgiu, Lampoltshammer, & Hofer, 2013; Hay, Castilla, Wulder, & Ruiz, 2005; Martha, Kerle, van Westen, Jetten, & Kumar, 2011). Human intervention necessary to provide domain knowledge during image analysis is replaced by knowledge formalization. A knowledge base is developed that delivers domain-specific information provided by experts for GEOBIA processes. In this context, ontology has emerged as a knowledge representation language, which represents and stores information in a format understandable to both humans and machines (Arvor, Durieux, Andrés, & Laporte, 2013; Belgiu, Hofer, & Hofmann, 2014).

The contribution of this study is to explore the need for automation processes in information extraction from remote sensing images. This exploration is further extended by developing a framework that can be applied in different case study examples. In a specific case study presented in detail here, we apply our framework to a land use/land cover (LULC) case study from Hobart, Tasmania, Australia. The key contributions of the framework are:

- The construction of a system-independent knowledge base necessary to carry out image analysis work. The knowledge creation process is separated from the image analysis to avoid subjectivity issues.
- The use of ontology to link low-level information extracted from image data with high-level domain knowledge from experts.
- The use of rule-based inference for image object classification.

In summary, we present an effort to bring automation to GEOBIA with minimum human intervention at the time of image classification by leveraging domain experts' knowledge using ontology.

In the following sections, we show how expert domain knowledge is formalized using an ontology and is applied in GEOBIA. We begin by providing a background to ontology and arguing the importance of ontology in the

GEOBIA process. We then review other research addressing the application of ontologies to GEOBIA. We report a case study that employs an ontological framework for GEOBIA. Finally, we conclude with a discussion of the application and potential benefits of using ontologies in GEOBIA.

The need for expert human involvement to extract meaningful information from remotely sensed image data limits automation in GEOBIA. In the next section, we explain how the ontology is used to address this limitation.

8.2 Ontology in GEOBIA

Ontology is not a new topic for geographic information science (GIS), and its use in remote sensing for advanced image analysis is steadily expanding (Agarwal, 2005; Arvor, Durieux, Andrés, & Laporte, 2013; Schuurman, 2006). The increasing availability of high-resolution earth observation data heightens the need to model these data as knowledge for data storage, interoperability, integration, and retrieval.

8.2.1 Definition

The classical definition of ontology is that it is a formal, explicit specification of a shared conceptualization (Gruber, 1993). Ontology is a machine-readable representation of a domain's terminology and the relationships among the terms in the domain. The ontology allows modeling of domain knowledge and provides a standardized vocabulary and rules to apply this vocabulary (Agarwal, 2005). Figure 8.1 is an example of an ontology constructed from a description of the plant genus Eucalyptus in natural language: Eucalyptus is a woody plant, found across Australasia, with maximum standing height ranging up to 90 m.

To better understand Gruber's definition of ontology, consider the example shown in Figure 8.1: Eucalyptus is *conceptualized* as an abstract entity; *explicitly* defined with constraints (vegetation: woody plant; standing height < 90 m; available region: Australia and Asia); *shared* with a consensual agreement between domain experts; and made *formal* by expressing these relationships in machine-readable language using knowledge representation (KR) languages (Studer, Benjamins, & Fensel, 1998).

Knowledge can be formally expressed using a KR language. Description logics (DLs), a family of KR language, can be used; in general, this is viewed as decidable fragments of first-order logics (Krötzsch, Simancik, & Horrocks, 2012). Domain knowledge is thus represented using different KR languages such as Resource Description Framework (RDF), Simple Knowledge Organization System (SKOS), or Web Ontology Language (OWL). RDF defines a data model to describe machine-understandable

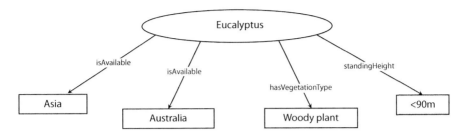

FIGURE 8.1
An example of an ontology.

semantics of data in terms of subject-predicate-object expression, which is commonly known as *triples* in RDF terminology (Broekstra et al., 2002). SKOS is the World Wide Web Consortium (W3C)–recommended data model and vocabulary to express knowledge organization systems (KOSs) such as thesauri and classification schemes (Baker et al., 2013). OWL is a language for modeling ontologies, which became a W3C recommendation in February 2004 (Bechhofer et al., 2004). The basic elements of OWL ontology are classes, individuals, and properties. Classes are sets of individuals and properties that exist either between individuals or between the object and a data type (Belgiu et al., 2013). In our work, we have used OWL language to develop an ontology necessary for LULC classification.

8.2.2 How GEOBIA Can Be Improved Using Ontology

There is an increasing use of GEOBIA methods in remote sensing image interpretation (Blaschke, 2010; Blaschke et al., 2014). However, there are barriers that limit the capacity to automate GEOBIA frameworks. The primary issues are semantic gaps, the need for human intervention, the uncertainty in rules transferability, and human cognitive bias. In this section, we will discuss how ontology helps to resolve these issues.

Semantic gap. A semantic gap is the lack of coincidence between low-level information extracted from image data and high-level information needed for semantic interpretation of images in extracting features (Smeulders, Worring, Santini, Gupta, & Jain, 2000). While GEOBIA makes use of spectral features to identify geo-objects in an image, this process requires human knowledge to interpret pixel values extracted from images into a meaningful feature. Low-level spectral information such as digital numbers (DN) for each band is extracted from satellite images. Extracting high-level semantic information from an image requires expert domain knowledge and implicit information contained in the context of the image region (Belgiu & Lampoltshammer, 2013; Bratasanu, Nedelcu, & Datcu, 2011; Zhang, Xu, Zhang, & Yu, 2014).

Knowledge representation techniques such as ontologies help to reduce semantic gaps (Andres, Arvor, & Pierkot, 2012). Initially, expert knowledge is formalized to develop a domain ontology. Next, image pixel values need to be converted into meaningful physical measurements, which are defined in the measurement ontology. The gap is reduced by making a semantic link between domain ontology and measurement ontology. Domain ontology describes the concepts, relations, and properties relevant to a generic domain (such as LULC). Measurement ontology describes the concepts and properties specified to define measurement values (such as High, Low). For example, an expert may provide knowledge that trees are high, while shrubs are low. We could define concepts such as *trees, shrubs, high,* and *low* and properties such as *standingHeight, hasMinThresholdValue.* Based on expert knowledge, we could relate <Trees> *standingHeight*<High> and <High> *hasMinThreshold-Value* <50m>. The values inside "<>" represent concepts (<Trees>, <High>) or literal values (<50m>) depending on the type of properties (object type or data type) used. In this way, we identify all the instances of concept "Trees" that have elevation value greater than 50 m.

Human intervention. One of the challenging problems identified in GEOBIA is that it requires human intervention in the form of expert knowledge for the reliable identification of remotely sensed image objects. Defining an effective segmentation algorithm involves a trial and error process, using human input to adjust the processing parameters. Thus, the segmentation process is ill-posed, with no unique solution, and different interpreters will delineate in different ways (Hay & Castilla, 2008). Reliable image interpretation depends on expert knowledge of the domain area (Arvor, Saint-Geours, Dupuy, Andrés, & Durieux, 2013). Experiments conducted to test the variability of rule-based classification on the same image analyzed by different interpreters showed that classification accuracy is directly influenced by subjective decisions being made by the interpreters (Belgiu, Drăguţ, & Strobl, 2014). Automation in image analysis is challenging, as this normally employs human interaction to analyze highly complex image contents (Hofmann et al., 2015). This leads to the need to formalize expert knowledge using knowledge representation techniques such as ontologies. The knowledge is formalized prior to image analysis with a mutual consensus of domain experts. The image interpretation task can then be based on the rules extracted from the knowledge base, so no human interaction is involved, thus eliminating the influence of a single individual and a single incidence in the image analysis process.

Transferability. Transferability of knowledge-based classification rules shows the possibility for automation in the object identification process but also raises concerns about decreasing classification accuracy (Leukert, Darwish, & Reinhardt, 2004). The reuse of rule sets becomes inapplicable when the region characterized by atmospheric differences and sensor types are different. The developed rule sets can be applied to other regions captured from the same sensor, but they may not be effective to use with different sensor types. Belgiu, Drăguţ, and Strobl's (2014) transferability assessment showed consistent results when rule sets were applied to a larger area from the same Worldview 2 satellite image. Tiede, Lang, Hölbling, and Füreder (2010) found that the transferability of rule sets developed for QuickBird images required slight adaptation when applied to GeoEye-1 images. The adaptation from the original rule sets was performed by visual inspection to modify spectral thresholds for classification of dwelling and vegetative classes. The changes in the rule set involved only the adjustment in the threshold parameter values to fit with new datasets. The need for adaptation in rule set parameters has also been identified in other transferability studies in such diverse applications as slum identification (Kohli, Warwadekar, Kerle, Sliuzas, & Stein, 2013), urban area classification (Tathiri, Shafri, & Hamedianfar, 2014), and glacial cirques extraction (Anders, Seijmonsbergen, & Bouten, 2015). Using generic knowledge models, Novack, Kux, Feitosa, and Costa (2014) showed the feasibility of rule set transfers between different image analysis systems.

Redefining the rule sets for every new image set or system is time-consuming and considered as a major obstacle in automating the image classification process. Expert knowledge plays a significant role in the design of the rule set, and such rule sets are transferable to areas with similar morphological structure and sensors (Anders et al., 2015). Ontology formalizes expert knowledge in developing rule sets independently of data. The generated rules are transferable, as these are generic. These generic rules can be localized to adapt rule sets specific to new study areas and data. Kohli et al. (2013) created generic slum ontology for image-based slum identification, which was later localized to classify other slum areas. The generic rule sets developed described the image class, and the threshold values were determined for every new set of data. For different areas and sensor types used, the rule sets require minimal changes, so there is a necessity for an automatic approach for calculating optimum threshold without user involvement to enter new values with a change in datasets (Tathiri et al., 2014).

Human cognitive bias. Interpretation of remote sensing images requires that interpreters have deep a priori domain knowledge. When there is human input in image analysis, the interpretation work may be biased or inadequate due to the limited expertise of individual experts (Belgiu et al., 2014). The subjectivity issues in interpretation could be improved if domain knowledge was created with the consensus of domain experts. Creating an ontology based on expert knowledge and using formalized knowledge at the time of image interpretation help to resolve the cognitive bias issues. This type of ontology, referred to as a *domain ontology*, is a knowledge on the specific domain as agreed by the experts (Agarwal, 2005; Guarino, 1998).

8.3 Previous Work

Ontologies have been applied to GEOBIA in a variety of case studies, which include the identification of urban and peri-urban landscapes (Durand et al., 2007), LULC classification (Andres et al., 2012; Belgiu et al., 2014), the identification of slum areas (Kohli, Sliuzas, Kerle, & Stein, 2012), and the classification of ocean satellite images (Almendros-Jimenez, Domene, & Piedra-Fernandez, 2013). In this section, we will discuss and present methodologies developed to exploit ontology-based competencies for image interpretation in a GEOBIA framework.

Classification in object-based image analysis is carried out using domain knowledge. The formalization of knowledge provides an insightful and systematic approach to accessing the information necessary to identify image objects. An ontology defines concepts and their relationships for knowledge representation and reasoning. In the context of remote sensing images, ontology helps to conceptualize image concepts (e.g., land, forest, water bodies, etc.) based on spectral, spatial, and textural features and link them to expert knowledge. A framework is proposed by Andres et al. (2012), where an image ontology is developed to define the core concepts of the image. Formally defined image concepts are used to describe segmented images with low-level features and to build an ontology that represents high-level concepts in the images to classify image segments. The need for formalizing expert knowledge for the automatic semantic interpretation of remote sensing images is highlighted in the study by Andrés, Pierkot, and Arvor (2013).

An ontology-based object recognition algorithm is proposed by Durand et al. (2007). This semantically matches concepts defined in an ontology with segmented image regions from a QuickBird image. This study also points out that knowledge formalization helps to bridge the semantic gap between expert knowledge and image information. The ontology-based modeling of semantic concepts relates the high-level and low-level features to reduce the semantic gap as, for example, demonstrated in ocean satellite image classification (Almendros-Jimenez et al., 2013). This rule-based ontological framework uses labeling and training rules to map low-level features and human expert rules to map high-level concepts of oceanic structures.

The feasibility of implementation of an ontology-based GEOBIA framework is undermined by the fundamental problem of integrating ontology into already existing GEOBIA systems. A method to couple ontologies in GEOBIA was proposed by Belgiu et al. (2014). The coupling was achieved by transforming the concept in the hierarchy as defined in the ontology into the class hierarchy needed by eCognition software using extensible style sheet language transformations (XSLT). The proposed framework highlighted that the strength of an integration approach is that it can ease the image analysis procedure and reduce its subjectivity.

Ontologies provide formalized knowledge, and with its integration into GEOBIA, the necessary domain knowledge required for the image interpretation process is achieved. To attain more advancement in GEOBIA, formalized knowledge needs to be reused in integrating images from different regions, from varying contexts, and across multiple systems. The transferability of rule sets for image interpretation is yet another vital research area in GEOBIA. An ontological framework for object-oriented image analysis to identify slum areas in remote sensing imagery was proposed, whereby a generic slum ontology was created with well-defined attributes characterizing slums (Kohli et al., 2012). The developed ontology was based on the knowledge obtained from 50 domain experts from 16 different countries. The generic slum ontology was locally adapted to guide the object-oriented image analysis (OOA) method for slum classification. In their subsequent work, the tests showed the transferability of an OOA method developed on one image area to other image areas (Kohli et al., 2013). The study concluded that the use of generic ontology serves as a comprehensive knowledge base, but still, a local adaptation of it is required to match the local conditions. This study ensured the applicability of the framework to different slum areas with different characteristics.

From the existing works, it is clear that GEOBIA requires expert domain knowledge for accurate image analysis, and that this knowledge can be incorporated into GEOBIA using ontologies. An ontology reduces the semantic gap between low- and high-level information used in the image interpretation process and improves the transferability of knowledge and methodology applied to one region into a different region.

8.4 Methodology

8.4.1 Ontological Framework for GEOBIA

The methodological approach is broadly categorized into four steps: (1) ontology construction, (2) segmentation and feature extraction, (3) rule sets development, and (4) image object classification. Figure 8.2 outlines the overall workflow for an ontology-based image classification, where the ontology construction process is independent of the segmentation and feature extraction module and linked to the classification module via a rule sets development module.

Ontology construction. Ontology construction is the first step, where a knowledge base is developed by domain experts. In this process, different classes from the selected domain are identified and arranged in a taxonomic hierarchy. The classes in taxonomy are defined with characteristic features that help to identify them and differentiate them from each other. Based on these identified features, different ontological properties are created, which are used to assign values to the respective class. As shown in Figure 8.2, this module independently creates knowledge that is necessary for developing classification rules.

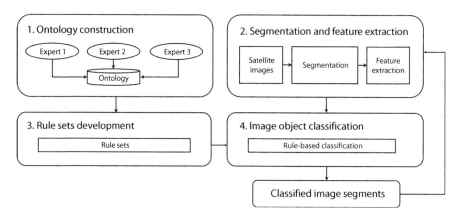

FIGURE 8.2
Workflow for ontology-driven image classification.

Segmentation and feature extraction. In the next step, the image is segmented using a segmentation algorithm. For each image segment, associated feature values are extracted and exported as comma separated values (CSV) files. These image segments are possible instances of different classes, and the extracted features are values that will be associated with the class using different data type properties defined in the ontology. To integrate these image segments with feature values into the ontology, a transformation of data from CSV format to a machine-understandable language such as RDF is necessary. Finally, the ontology schema and transformed RDF instances are merged together and input to the classification module as depicted in Figure 8.2.

Developing the rule sets. Rule sets are developed by exploiting the feature values of each image object instance to classify them into LULC classes. The rules are created using the relational operator (e.g., greater than, less than) along with threshold values for each feature. The selection of features and its threshold value for each class are determined by the expert's experience and knowledge. The rules are written using Semantic Web Rule Language (SWRL*), as recommended by W3C. Figure 8.2 shows that this module makes use of ontology to develop rules, which are exported to the classification module.

Image object classification. Finally, image segment instances are assigned to their respective classes based on the defined rule sets using an inference engine. Several inference engines are available, such as Pellet,[†] Hermit,[‡] and KOAN2.[§] GEOBIA is an iterative process in which there is multiple feedback loop between segmentation and classification processes; they interact as

---

* www.w3.org/Submission/SWRL/
[†] http://github.com/clarkparsia/pellet
[‡] www.hermit-reasoner.com/
[§] http://kaon2.semanticweb.org

presented in Figure 8.2. The classification results become an input for next-level segmentation.

8.4.2 Case Study: Identifying LULC Geo-Objects in Satellite Imagery

Study area. The study area selected for this case study is situated in the city of Hobart in Tasmania, Australia. Figure 8.3 shows the satellite imagery of the study area projected with Universal Transverse Mercator/Geocentric Datum of Australia (UTM/GDA) transverse Mercator projection. Although the total area of the study site is relatively small (0.43 sq. km.), it contains all the desired LULC categories, such as Water, Land, Shadow, Urban, Vegetation, Trees, and Grassland, defined in the ontology.

Data. QuickBird imagery of 11-bit radiometric resolution and 2.44 m multispectral resolution with 0.61 m panchromatic and four spectral bands, blue (0.45–0.52 μm), green (0.52–0.60 μm), red (0.63–0.69 μm) and near-infrared (NIR) (0.76–0.90 μm), was used. The spectral reflectance data were used to calculate indices such as normalized difference vegetation index (NDVI) and normalized difference water index (NDWI). NDVI in Equation 8.1 is calculated using red and NIR bands where the vegetation patch absorbs the visible spectrum (red band) and reflects the near-infrared band (Rouse, Haas, Schell, & Deering, 1974). NDWI in Equation 8.2 is calculated using the reflected NIR and green bands to extract water bodies and eliminate the presence of soil and terrestrial vegetation features (Gao, 1996; McFeeters, 1996).

$$NDVI = \frac{NIR - RED}{NIR + RED} \tag{8.1}$$

$$NDWI = \frac{Green - NIR}{Green + NIR} \tag{8.2}$$

Methodology. The key objective of this case study was to demonstrate the use of ontology to create rule sets necessary for image object classification. As described in the methodology, an LULC ontology was constructed as shown in Figure 8.4. The whole ontology development process was carried out using Protégé,* an open-source ontology editing tool. A similar class hierarchy was developed in eCognition software for comparison between the ontological framework and eCognition rule-based classification.

The satellite image was segmented using multiresolution segmentation in the eCognition software with the following parameters: scale = 50, shape = 0.5, compactness = 0.5, image layer weight: blue = 1, green = 1, NIR = 2, red = 1. The segmentation produced 1024 image objects, as shown in Figure 8.5. Low-level

---

* http://protege.stanford.edu/

FIGURE 8.3
Study area showing the heterogeneous LULC cover of part of Hobart city, Tasmania, Australia.

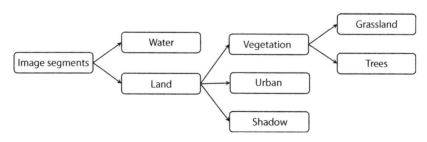

FIGURE 8.4
Land use/land cover (LULC) ontology.

feature (e.g., NDVI, NDWI, mean NIR, or standard deviation NIR) values for segmented image objects were exported as a CSV file. Each row of the CSV file represents individual image segments with different feature values. Data were transformed from CSV to RDF format using OpenRefine (formerly known as GoogleRefine) and the RDF refine tool, which is an extension to OpenRefine for exporting data in RDF format. Later, Protégé software was used to merge the ontology schema and RDF instances.

For each class, rules were developed comprising relevant features with appropriate threshold values. Figure 8.6 shows the rules written in SWRL, which were executed using the semantic reasoner. Pellet, an OWL reasoner installed as a plugin in Protégé software, executes the rules and assigns the entire image segment instance into respective LULC classes.

FIGURE 8.5
Multiresolution segmentation result from eCognition software.

ImageSegments(?x) ^ hasMeanNir(?x, ?z) ^ swrlb:greaterThanOrEqual(?z, "42.53"^^xsd:double) → Land(?x)

ImageSegments(?x) ^ hasNDWI(?x, ?z) ^ swrlb:greaterThanOrEqual(?y, "–0.23"^^xsd:double) ^ hasMeanNir(?x, ?z) ^ swrlb:lessThan(?z, "42.53"^^xsd:double) → Water(?x)

Land(?x) ^ hasMeanBrightness(?x, ?y) ^ swrlb:lessThanOrEqual(?y, "45.0"^^xsd:double) → Shadow(?x)

Land(?x) ^ hasNDVI(?x, ?y) ^ swrlb:lessThanOrEqual(?y, "0.18"^^xsd:double) → Urban(?x)

Land(?x) ^ hasNDVI(?x, ?y) ^ swrlb:greaterThan(?y, "0.18"^^xsd:double) → Vegetation(?x)

Vegetation(?x) ^ hasSDNir(?x, ?y) ^ swrlb:lessThan(?y, "20.0"^^xsd:double) → Grassland(?x)

Vegetation(?x) ^ hasSDNir(?x, ?y) ^ swrlb:greaterThan(?y, "20.0"^^xsd:double) → Trees(?x)

FIGURE 8.6
Semantic Web Rule Language to classify different LULC classes.

Results. The outcome of the execution of the rules is shown in Table 8.1, representing a hierarchical output. The result showed 106 Water and 918 Land objects out of 1024 segmented image objects in Level 1 hierarchy. Land objects are further classified as 149 Shadow, 208 Urban, and 561 Vegetation in Level 2 hierarchy. The vegetation class is further classified as 211 Grassland and 350 Trees objects in Level 3 hierarchy. The classification results (carried out at three different levels) are shown in Figure 8.7. The first-level classification results in Water and Land objects. This provides an input for the second level, where only those land areas are re-segmented and classified into Shadow, Urban, and Vegetation objects.

To check the robustness of the ontology-based image object classifier, the output was compared with the rule-based classifier available in eCognition software. The class hierarchy defined in the ontology was created in eCognition. SWRL rules were rewritten in eCognition, and the classification was carried out. The ontology-based image classification and the eCognition rule-based classification produced the same results in terms of the area covered by classified objects, as depicted in Figure 8.8. This demonstrates that the ontological framework serves as a complementary solution for image object classification in GEOBIA. The need for ontological rule-based classification is justified with the availability of a

TABLE 8.1

Result of Ontology-Based Image Classification

Class	Number of Classified Image Segments	Total
Water	106	1024
Land	918	
Shadow	149	
Urban	208	918
Vegetation	561	
Grassland	211	561
Trees	350	

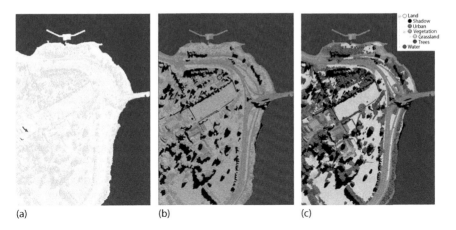

(a)   (b)   (c)

FIGURE 8.7
Classification result at different levels: (a) Level 1 (Water, Land); (b) Level 2 (Water, Vegetation, Shadow, and Urban); (c) Level 3 (Water, Trees, Grassland, Shadow, and Urban).

standardized and streamlined knowledge creation process and its use in GEOBIA. Further, the use of standard languages such as OWL and SWRL permits semantic interoperability and reasoning. Along with classification, the ontology-based approach has additional advantages, which are discussed in the next section.

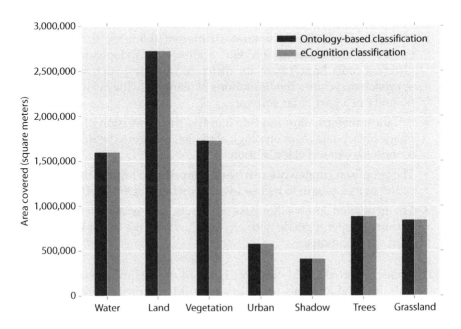

FIGURE 8.8
Area comparison of each classified class between eCognition software and ontological classification.

8.5 Discussion and Conclusions

The potential of employing ontological frameworks has been recognized by the experts working in GEOBIA because of the importance of knowledge in the image object identification process and its influence on the output. The knowledge converging from various disciplines needs to be mutually agreed by domain experts; it needs to be contextually aware and easily transferable. Domain knowledge is broad and diverse in nature, and it will rarely be held in its entirety by an individual interpreter. Thus, an ontology-based framework has the potential to transform existing GEOBIA into automated knowledge-based systems by converging consensual knowledge from experts beforehand and reducing or eliminating the involvement of experts at the time of analysis. In the framework, an ontology construction module is loosely coupled with an image classification module. Such isolation of domain knowledge construction allows domain experts to develop and enrich a knowledge base independently without understanding the segmentation and classification techniques or the influence of a targeted application area. In addition, the framework allows the use of collaborative knowledge management tools to keep the knowledge base up to date with a consensus of domain experts.

From the application perspective, the key benefits of using an ontological framework are:

- The framework clearly separates different modules, such as segmentation and classification, which allows the independent use of different tools or software for different modules. Such a modular approach overcomes the limitations of using only the available functionality of a particular software.
- A data transformation module handles the conversion of extracted image object data to an ontological model, that is, to the RDF format, for ontology-based classification.
- The proposed framework can be developed into a standalone application or as a plugin to be used with existing GEOBIA software.
- The generated knowledge base, rule sets, and image classification output are interoperable with any systems with the use of W3C standard specifications.

The work reported in this chapter is based on preliminary research but covers all the fundamentals necessary to build a knowledge-driven object-based image analysis. Our study is a proof of concept that demonstrates the feasibility of using an ontology for GEOBIA. The framework proposed in this study brings automation into meaningful information extraction from imagery, which has been long desired by the remote sensing community.

In summary, GEOBIA has emerged as a new paradigm in the remote sensing domain that requires expert knowledge for precise image object interpretation. From this perspective, ontology formalizes necessary human knowledge and thus helps to automate the GEOBIA process by limiting human intervention. Apart from this, ontology assists in tackling human cognitive bias and increases the applicability of GEOBIA processes, enabling the transferability of knowledge and rule sets.

References

Agarwal, P. (2005). Ontological considerations in GIScience. *International Journal of Geographical Information Science*, 19, 501–536.

Almendros-Jimenez, J. M., Domene, L., & Piedra-Fernandez, J. A. (2013). A framework for ocean satellite image classification based on ontologies. *IEEE Journal of Selected Topics in Applied Earth Observations and Remote Sensing*, 6(2), 1048–1063.

Andrés, N. S., Seijmonsbergen, A. C., & Bouten, W. (2015). Rule set transferability for object-based feature extraction: An example for cirque mapping. *Photogrammetric Engineering and Remote Sensing*, 81(6), 507–514.

Andrés, S., Arvor, D., & Pierkot, C. (2012, 25–29 November). Towards an ontological approach for classifying remote sensing images. In *Proceedings of the 2012 Eighth International Conference on Signal Image Technology and Internet Based Systems* (pp. 825–832). Naples, Italy: IEEE.

Andrés, S., Pierkot, C., & Arvor, D. (2013, 24 February–1 March). Towards a semantic interpretation of satellite images by using spatial relations defined in geographic standards. In *Proceedings of the Fifth International Conference on Advanced Geographic Information Systems, Applications, and Services* (pp. 99–104). Nice, France: IARIA.

Aplin, P., & Smith, G. (2008, 3–11 July). Advances in object-based image classification. In C. Jun, J. Jie & J. V. Genderen (Eds.), *Proceedings of the XXIst ISPRS Congress* (pp. 725–728). Beijing, China: ISPRS.

Argyridis, A., & Argialas, D. P. (2015). A fuzzy spatial reasoner for multi-scale GEOBIA ontologies. *Photogrammetric Engineering and Remote Sensing*, 81(6), 491–498.

Arvor, D., Durieux, L., Andrés, S., & Laporte, M. A. (2013). Advances in geographic object-based image analysis with ontologies: A review of main contributions and limitations from a remote sensing perspective. *ISPRS Journal of Photogrammetry and Remote Sensing*, 82, 125–137.

Arvor, D., Saint-Geours, N., Dupuy, S., Andrés, S., & Durieux, L. (2013, 13 April). Identifying optimal classification rules for geographic object-based image analysis. In *Proceedings of the XVI Simpósio Brasileiro de Sensoriamento Remoto* (pp. 2290–2297). Igauçu, Brazil: INPE.

Baker, T., Bechhofer, S., Isaac, A., Miles, A., Schreiber, G., & Summers, E. (2013). Key choices in the design of Simple Knowledge Organization System (SKOS). *Journal of Web Semantics*, 20, 35–49.

Bechhofer, S., Harmelen, F. V., Hendler, J., Horrocks, I., McGuinness, D. L., Patel-Schneider, P. F., & Stein, L. A. (2004, 10 February 2004). OWL Web Ontology Language Reference. Retrieved from www.w3.org/TR/owl-ref/.

Belgiu, M., Drăguţ, L., & Strobl, J. (2014). Quantitative evaluation of variations in rule-based classifications of land cover in urban neighbourhoods using WorldView-2 imagery. *ISPRS Journal of Photogrammetry and Remote Sensing*, 87(100), 205–215.

Belgiu, M., Hofer, B., & Hofmann, P. (2014). Coupling formalized knowledge bases with object-based image analysis. *Remote Sensing Letters*, 5(6), 530–538.

Belgiu, M., & Lampoltshammer, T. J. (2013). Ontology-based interpretation of very high resolution imageries–grounding ontologies on visual interpretation keys. In *Proceedings of the 16th AGILE Conference on Geographic Information Science* (pp. 1–5). Leuven, Belgium: AGILE .

Belgiu, M., Lampoltshammer, T. J., & Hofer, B. (2013). An extension of an ontology-based land cover designation approach for fuzzy rules. In T. Jekel, A. Car, J. Strobl & G. Griesebner (Eds.), *Gi_Forum 2013: Creating the Gisociety* (pp. 59–70). Salzburg: Austrian Academy of Sciences Press.

Benz, U. C., Hofmann, P., Willhauck, G., Lingenfelder, I., & Heynen, M. (2004). Multi-resolution, object-oriented fuzzy analysis of remote sensing data for GIS-ready information. *ISPRS Journal of Photogrammetry and Remote Sensing*, 58, 239–258.

Blaschke, T. (2010). Object-based image analysis for remote sensing. *ISPRS Journal of Photogrammetry and Remote Sensing*, 65(1), 2–16.

Blaschke, T., Hay, G. J., Kelly, M., Lang, S., Hofmann, P., Addink, E., & Tiede, D. (2014). Geographic object-based image analysis—towards a new paradigm. *ISPRS Journal of Photogrammetry and Remote Sensing*, 87(100), 180–191.

Blaschke, T., Lang, S., Lorup, E., Strobl, J., & Zeil, P. (2000). Object-oriented image processing in an integrated GIS/remote sensing environment and perspectives for environmental applications. *Environmental Information for Planning, Politics and the Public*, 2, 555–570.

Blaschke, T., & Strobl, J. (2001). What's wrong with pixels? Some recent developments interfacing remote sensing and GIS. *Geo-Informations-Systeme*, 14(6), 12–17.

Bratasanu, D., Nedelcu, I., & Datcu, M. (2011). Bridging the semantic gap for satellite image annotation and automatic mapping applications. *IEEE Journal of Selected Topics in Applied Earth Observations and Remote Sensing*, 4(1), 193–204.

Broekstra, J., Klein, M., Decker, S., Fensel, D., van Harmelen, F., & Horrocks, I. (2002). Enabling knowledge representation on the Web by extending RDF schema. *Computer Networks*, 39(5), 609–634.

DigitalGlobe. (2014). WorldView-3 Data Sheet. Retrieved from https://www.spaceim-agingme.com/downloads/sensors/datasheets/DG_WorldView3_DS_2014.pdf.

Durand, N., Derivaux, S., Forestier, G., Wemmert, C., Gancarski, P., Boussaid, O., & Puissant, A. (2007, 29–31 October). Ontology-based object recognition for remote sensing image interpretation. In *Proceedings of the 19th IEEE International Conference on Tools with Artificial Intelligence* (pp. 472–479). Patras: IEEE.

Gao, B.-C. (1996). NDWI—A normalized difference water index for remote sensing of vegetation liquid water from space. *Remote Sensing of Environment*, 58(3), 257–266.

Gruber, T. R. (1993). A translation approach to portable ontology specifications. *Knowledge Acquisition*, 5, 199–220.

Guarino, N. (1998). Formal ontology and information systems. In N. Guarino (Ed.), *Proceedings of the International Conference on Formal Ontology in Information Systems (FOIS1998)* (pp. 3–15). Trento, Italy: IOS Press.

Hay, G. J., & Castilla, G. (2008). Geographic Object-Based Image Analysis (GEOBIA): A new name for a new discipline. In T. Blaschke, S. Lang & G. J. Hay (Eds.), *Object-based Image Analysis: Spatial Concepts for Knowledge-driven Remote Sensing Applications* (pp. 75–89). Berlin, Heidelberg: Springer.

Hay, G. J., Castilla, G., Wulder, M. A., & Ruiz, J. R. (2005). An automated object-based approach for the multiscale image segmentation of forest scenes. *International Journal of Applied Earth Observation and Geoinformation*, 7(4), 339–359.

Hofmann, P., Lettmayer, P., Blaschke, T., Belgiu, M., Wegenkittl, S., Graf, R., …, Andrejchenko, V. (2015). Towards a framework for agent-based image analysis of remote-sensing data. *International Journal of Image and Data Fusion*, 6(2), 115–137.

Kohli, D., Sliuzas, R., Kerle, N., & Stein, A. (2012). An ontology of slums for image-based classification. *Computers, Environment and Urban Systems*, 36(2), 154–163.

Kohli, D., Warwadekar, P., Kerle, N., Sliuzas, R., & Stein, A. (2013). Transferability of object-oriented image analysis methods for slum identification. *Remote Sensing*, 5(9), 4209–4228.

Krötzsch, M., Simancik, F., & Horrocks, I. (2012). A description logic primer. arXiv preprint arXiv:1201.4089.

Lauer, D. T., Morain, S. A., & Salomonson, V. V. (1997). The landsat program: Its origins, evolution, and impacts. *Photogrammetric Engineering and Remote Sensing*, 63(7), 831–838.

Leukert, K., Darwish, A., & Reinhardt, W. (2004). Transferability of knowledge-based classification rules. In O. Altan (Ed.), *Proceedings of the XXth ISPRS Congress* (pp. 1059–1064). Istanbul, Turkey: ISPRS.

Martha, T. R., Kerle, N., van Westen, C. J., Jetten, V., & Kumar, K. V. (2011). Segment optimization and data-driven thresholding for knowledge-based landslide detection by object-based image analysis. *IEEE Transactions on Geoscience and Remote Sensing*, 49(12), 4928–4943.

McFeeters, S. K. (1996). The use of the Normalized Difference Water Index (NDWI) in the delineation of open water features. *International Journal of Remote Sensing*, 17(7), 1425–1432.

Novack, T., Kux, H., Feitosa, R. Q., & Costa, G. A. O. P. (2014). A knowledge-based, transferable approach for block-based urban land-use classification. *International Journal of Remote Sensing*, 35(13), 4739–4757.

Rouse Jr, J. W., Haas, R., Schell, J., & Deering, D. (1974). Monitoring vegetation systems in the Great Plains with ERTS. *NASA Special Publication*, 351, 309.

Schuurman, N. (2006). Formalization matters: Critical GIS and ontology research. *Annals of the Association of American Geographers*, 96(4), 726–739.

Smeulders, A. W. M., Worring, M., Santini, S., Gupta, A., & Jain, R. (2000). Content-based image retrieval at the end of the early years. *IEEE Transactions on Pattern Analysis and Machine Intelligence*, 22(12), 1349–1380.

Studer, R., Benjamins, V. R., & Fensel, D. (1998). Knowledge engineering: Principles and methods. *Data and Knowledge Engineering*, 25, 161–197.

Tathiri, A., Shafri, H. Z. M., & Hamedianfar, A. (2014, 13–14 November). Development of transferable rule-sets for urban areas using QuickBird satellite imagery. In *Proceedings of the IEEE International Conference on Aerospace Electronics and Remote Sensing Technology* (pp. 229–233). Yogyakarta: IEEE.

Tiede, D., Lang, S., Hölbling, D., & Füreder, P. (2010, 29 June–2 July). Transferability of OBIA rulesets for IDP Camp Analysis in Darfur. In E. A. Addink & F. M. B. Van Coillie (Eds.), *Proceedings of the GEOBIA 2010: Geographic Object-Based Image Analysis* (pp. 1–6). Ghent, Belgium: ISPRS.

Weih Jr, R. C., & Riggan Jr, N. D. (2010). Object-based classification vs. pixel-based classification: comparative importance of multi-resolution imagery. In E. A. Addink & F. M. B. Van Coillie (Eds.), *Proceedings of the GEOBIA 2010: Geographic Object-Based Image Analysis* (pp. 1–6). Ghent, Belgium: ISPRS.

Zhang, L., Xu, T., Zhang, J., & Yu, Y. (2014, 19–21 August). A knowledge-based procedure for remote sensing image classification. In *Proceedings of the 11th International Conference on Fuzzy Systems and Knowledge Discovery* (pp. 72–76). Xiamen: IEEE.

Index

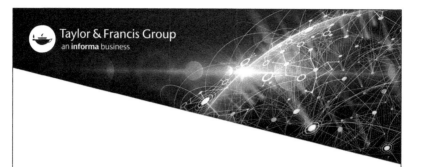